SOUTH STAFFORDSHIRE IRONMASTERS

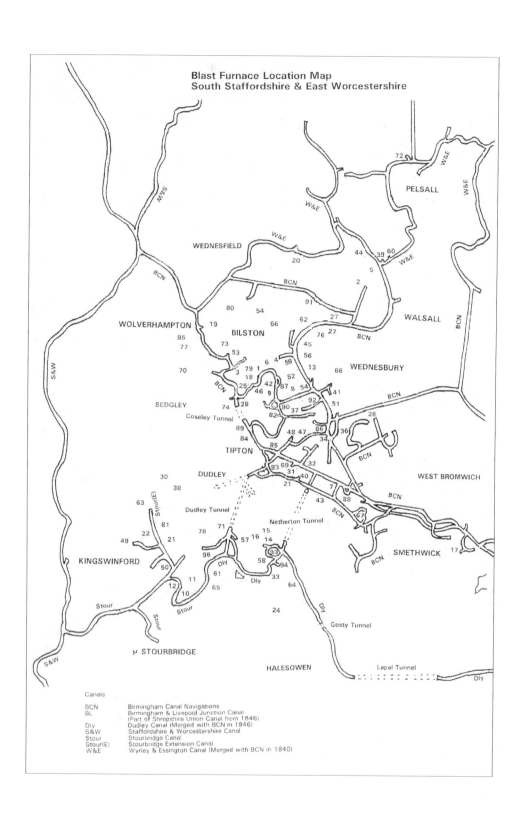

Blast Furnace Location Map
South Staffordshire & East Worcestershire

SOUTH STAFFORDSHIRE IRONMASTERS

RAY SHILL

The
History
Press

Frontispiece: Blast furnace location map. The numbers on the map correspond with those in Appendix Three (page 182).

First published 2008
Reprinted 2020

The History Press
97 St George's Place, Cheltenham,
Gloucestershire, GL50 3QB
www.thehistorypress.co.uk

British Library Cataloguing in Publication Data.
A catalogue record for this book is available from the British Library.

ISBN 978 0 7524 4831 2

Typesetting and origination by The History Press
Printed in Great Britain by TJ International Ltd, Padstow, Cornwall.

CONTENTS

INTRODUCTION

With each generation that passes, aspects of our past are forgotten and several generations have now gone by since ironmaking and working was an important trade in the Black Country. The people who worked in this occupation are also generally forgotten. They were drawn to the region from different parts of England, Ireland, Scotland and Wales, in search of work. Integral to the trade were the employers who organised the industry; these were a mixed bunch of people from different backgrounds who came to the district with a common cause: the making of iron. Those in charge were known as the 'ironmasters' and they were the owners of the blast furnaces that smelted the iron from the ore, or were the proprietors of ironworks where the iron was then made malleable. It was a trade conducted throughout Britain in areas where iron ore, or ironstone, was found. Within the West Midlands ironstone was found in parts of Shropshire, Staffordshire, Warwickshire and Worcestershire.

Shropshire is well known for its associations with the iron trade and the Ironbridge Gorge has been called the 'cradle of the iron industry'. While the area around the Ironbridge Gorge has been designated a World Heritage Site and the iron industry there has been especially commemorated in the museums around the Gorge, much less has been done to recognise the contribution of the Black Country ironmasters of South Staffordshire and East Worcestershire.

A champion of the Black Country cause is the Black Country Museum, but even this museum has not tackled the subject in depth and has done little to improve understanding of the formative years that essentially created the Black Country. Universities and colleges might also have done more, but perhaps both have been handicapped by constantly changing trends that have accompanied a serious decline in industry and the reallocation of land for housing, shops or warehouses.

The iron industry in the Black Country was as old, or older, than that in Shropshire; it developed through the use of watermills, where the power of local rivers was harnessed. The use of watermills for this purpose can arguably be seen as the first Industrial Revolution. The innovation of steam power for pumping water and draining mines that was eventually refined for driving machinery was the next important stage. It was during the mid- to late eighteenth century that this all-important step happened and it was accompanied by the establishment of a national network of canals and so is seen by many as the time of the Industrial Revolution. If importance is attached to the role of the watermill, however, then perhaps the development of canals and steam power belongs to a second Industrial Revolution. Taking this analogy further, other significant 'revolutions' could include the discovery of electricity, the development of plastics and, in more recent times, the silicon chip.

Ironmaking in the Black Country began long before the term 'Black Country' was coined. It was started by itinerant bloomers, who moved their bloomeries around the district to make use of local supplies of ore, smelting it with charcoal made from wood of the forest. Water-powered bloomeries on permanent sites were the next step and these were replaced in turn by the blast furnace. Charcoal-fuelled blast furnaces were used well into the second half of the eighteenth century. These were, in turn, replaced by coke-fired smelting furnaces whose use was pioneered in this district by the Wilkinson family.

The Earl of Dudley's Round Oak Ironworks was one of the most carefully laid out ironmaking works of its day. Completed in 1857, the puddling furnaces and rolling mills were arranged in an orderly fashion and linked by standard-gauge and narrow-gauge railways. The former provided transport of coal to the furnaces and mills, and moved the finished iron away for distribution. The narrow-gauge track brought pig iron from the New Level Furnaces or the canal-side pig yard. (*Dudley Archives*)

Coke-smelted iron had a long period of development and was made necessary through dwindling supplies of charcoal. Once perfected, however, this process had the capacity for further improvement: producing metallic iron in greater amounts. While the Shropshire ironmasters working around Ironbridge can justly take the credit for the successful development of coke smelting and despite the fact that the mines about them had good-quality ironstone to make high-quality iron, the iron trade around South Staffordshire developed to a far greater extent and eventually became the greatest producer of finished iron in the country.

Those who have recorded the rise of the Black Country iron trade might have been both more vocal and kinder in its support. Few have commented on the role of ironmasters such as the Wilkinsons, Firmstones, Parkers, Banks and Baldwins who brought ironmaking skills out of Shropshire. John Wilkinson, in particular, commenced ironmaking by the coke-smelting process at Bradley. He achieved this through a new patent for improved bellows for the blast furnace, invented by his father, Isaac Wilkinson.

While Shropshire ironmaking continued to improve, it was gradually outstripped by the Black Country trade. Black Country ironmasters had their share of success and failure, profits and loss, wealth and bankruptcy. Such is the nature of the trade that supply and demand created periods of expansion and then, through overproduction, an inevitable slump. Political factors also had an influence on the iron trade; wars, for example, created an increased demand for iron for ordnance. When the battles were over and the wars won or lost, the bigger losers were the ironmasters and their workforce.

Matters changed through the mid-nineteenth century, during the reign of Queen Victoria, when innovation and invention reached new heights and iron came to be used in a multitude of applications. With this increased demand came many more furnaces of modern design, often in places outside the Midlands. Nevertheless, the Black Country ironmasters played an important role in the improvement of ironmaking, both in furnace design and the working up of iron into a finished product.

Smelting iron from its ores was only part of the ironmaking process. The metal might be moulded into shapes, as cast iron, or worked up as malleable iron. It was the malleable iron that had the most varied uses as hoops, plates, strips, rods, sheets, strips and wire. The road to making

Canal transport was essential to the development of the local iron industry and continued to provide an integral service through to the twentieth century. The above view shows canal boats lined along the wharf belonging to the London Ironworks, Oldbury. These works were on the now disused Oldbury Loop near Whimsey Bridge. (*Sandwell Archives*)

malleable iron more efficiently proved just as difficult as the process which perfected iron smelting through the coke process. Eventually, a method was invented that was called 'puddling', and this process created a whole new type of worker that was employed to work the iron up. Several West Midlands ironmasters made contributions in this respect. Joseph Hall deserves the greatest credit for his improvements to the puddling process, which enabled South Staffordshire to excel in this trade.

This book investigates the main ironmaster families and their relationships with ironmaking throughout Britain. They were a varied group, drawn from different regions and different faiths. There was a strong nonconformist element, but as some gained social status family members sometimes converted to the Anglican faith.

The lives of those connected to ironmasters were drawn into the ups and down of the iron trade and suffered their greatest reversal after the slump of 1857. From that point onwards, iron furnaces came to decline in number, although finished ironworks continued to thrive. Finally, corporate investment took over the running of many of the surviving furnaces and ironworks. Directors made decisions influenced by the shareholders. It was a different world and one that marked the end of the independent ironmaster.

After the trade became corporate the workers also came to have an input, especially regarding pay and conditions, through their collective associations, or unions. These unions came into existence during the 1860s after years of abuse by some ironmasters. Those that withdrew their labour were harshly treated by local militia, who suppressed any demonstrations, and local magistrates who fined or handed out prison terms for any who left their place of employment without the permission of their employers. There was also the sinister 'truck system' that kept workers in their jobs through the debts incurred in the part payment of wages with goods or drink.

Abbreviations

BCN   Birmingham Canal Navigations
GWR   Great Western Railway
LNWR   London & North Western Railway
OWWR   Oxford, Worcester & Wolverhampton Railway
SSR   South Staffordshire Railway

Calendar Dates

This book deals with the events in ironmaking history from the Middle Ages onwards. Readers should be aware that there was change in the calendar from January 1752 when the Julian calendar was replaced with the Gregorian calendar. The Julian calendar year commenced on 25 March and ended on 24 March. The Gregorian calender's year commenced on 1 January and ended on 31 December. For the purpose of this book the dates are quoted as originally reported.

Transport Links

Transport was an essential part of the ironmaking process. Iron smelted at one location was commonly moved to another location to be worked up in a foundry to make cast iron, or through the finery and chafery, or at later times the puddling furnace to make bar iron. Bar, or wrought, iron was then taken to be worked up further into hoops, plates, rails, rods, sections or sheets and these would be sold on for making specific products.

The movement of iron in its various forms was conducted along the canals, railways and roads of the West Midlands.

Roads were originally maintained by the parish, but this practice fell into decline through the increase of commerce. A system of turnpikes came into existence whereby major roads were repaired and maintained. The cost of the venture, and any profit, was funded through tolls charged for vehicles that passed along it. South Staffordshire and East Worcestershire had a complex network of turnpikes that developed during the eighteenth and early nineteenth centuries.

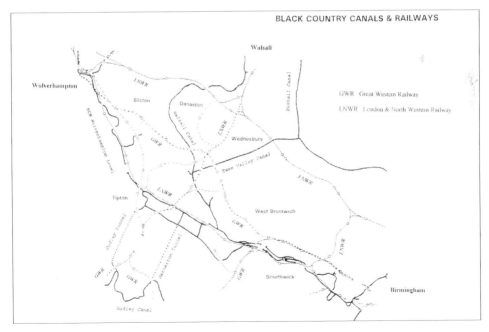

South Staffordshire smelting furnaces, forges and rolling mills benefitted from a comprehensive network of railways and canals. This map shows the extent of the network that was attained by 1900.

The most significant transport change was the building of the canal network. Ironmasters especially benefited from the construction of the Birmingham Canal Navigations (BCN) and the links it made with adjacent waterways; the Coventry, Dudley, Staffordshire and Worcestershire, Stourbridge, Stratford-upon-Avon, Warwick & Birmingham, Worcester & Birmingham and Wyrley & Essington canals.

Public railways reached the West Midlands in 1837 and a boom in construction that lasted from 1845–1860 ensured a very comprehensive local network of lines. Promoted initially by many different companies, ownership eventually coalesced into three concerns: the Great Western Railway (GWR), Midland Railway (MR) and London North Western Railway (LNWR).

Several ironmasters invested in the railway schemes. Their reasons were varied, some hoped to profit from the railway mania while others had a genuine interest in improving the supply of raw materials and the transport of finished goods. Not all schemes came to fruition. Such was the number of scheme proposals that Parliament appointed special committees to select the best options and weed out the duplicate, or non-practical, schemes.

Watermills

Watermills were utilised for various roles, the most common being the milling of grain. However, watermills were also adapted for a number of other uses including fulling and the working of brass, copper, iron or steel. From the sixteenth century, watermills came to have a crucial role in the ironmaking process. Water driving a waterwheel provided power for the blast engine, iron refining in the finery and chafery as well as the boring, rolling and slitting of iron.

Throughout Britain, streams and rivers were adapted to utilise the power of water through a process that took up a considerable portion of land. The skills of the millwright making the arrangement might involve the construction of artificial channels to divert water to the mill and a specially constructed pool to store water to assist the working of the mill. A mill leat, or race, was cut from the stream and a tail race was also cut to take the water back to the main stream. Where there was a regular supply of water and a fast flow, mills were placed adjacent to the river or stream and lacked the infrastructure of the mill pool and mill races. In the first case, the construction of the water courses involved additional cost and maintenance. The possibility of floods also had to be considered, therefore, additional bypass channels had to be cut with weirs or flood gates to divert water away from the mill pool.

The mills can be classified into four categories:

1. Mill with an external wheel built on the side and driven by the main stream or a leat.
2. Island mill: Floodgates sometimes conduct water away underground.
3. Mill with an internal wheel: Bypass water runs parallel to wheel
4. Straddle mill built over stream.

The arrangement of the waterwheel might also vary. Three common forms were:

1. Undershot: The wheel is fitted to use the flow of the river to drive it. Several mills on the Stour adopted this principal.
2. Breastshot: The water was directed onto the side of the wheel from a height.
3. Overshot: The water was directed on to the top of the wheel.

With iron mills, the power of the wheel was adapted through gearing to drive different equipment such as the finery, stamp or rolling mill. The mill could also employ subsidiary wheels, with tail races passing under the mill.

Each wheel was used to perform a different function that had a direct drive to a rolling mill or slitting mill. Alternatively, the wheel might turn a camshaft that worked the chafery, finery, blast furnace or tilt hammer.

GLOSSARY OF IRONWORKING TERMS

To aid the reader with understanding terms used in the text the following glossary has been compiled.

Reference: W.K.V. Gale, *The Iron and Steel Industry a Dictionary of Terms*, David & Charles (1971)

Air Furnace
A large size reverberatory furnace used for melting down cast iron before the invention of the cupola, and subsequently utilised for melting large pieces of scrap iron.

Ball Furnace
A furnace used for reheating wrought iron ready for rolling. It is also sometimes known as a 'mill furnace'. In a strict definition, the mill furnace was used for reheating the piles of iron before they proceeded to the rolling mill. The ball furnace was employed to heat balls of scrap iron.

Balling Furnace
This is a term which appears in early trade advertisements. It was probably another term for a puddling furnace.

Blast Furnace
This remains the primary producer of iron from iron ore, although its shape and design have been improved over the years. The basic principle involves a stream of air, or other gas, that is forced through openings, or tuyeres, into the furnace and aids combustion. The air could pass through the furnace as a cold blast or hot blast. The iron produced was drawn off in a molten state through a tap hole and collected in sand moulds. The iron, once cooled, was known as 'pig iron', and the pigs would either be used to make cast iron or malleable iron.

Bloomery
A small charcoal-fired hearth that was used to make wrought iron directly from the ore.

Bloom
This is an intermediate in the ironmaking process. The term describes a mass of wrought iron, produced through hammering, that is ready for further working, such as rolling.

Bulldog
Calcined, puddling furnace cinder, or slag, used as fettling.

Calcine
The act of concentrating iron ore by roasting, or burning, to drive off carbon dioxide and water and to oxidise the ore ready for smelting. The act of concentration was also of benefit when transporting the ore by rail or canal.

Cold Blast
Air forced through a blast furnace, which has not been previously heated.

Chafery
A small hearth that was fired with charcoal, coal or coke and used for heating iron blooms ready for reworking.

Chilled Roll
A rolling mill roll with a hard surface, used for finishing.

Cinder
A Black Country term for the slag produced from a blast furnace, puddling furnace or mill furnace.

Cinder Pig
Pig iron that is made from a mixture of iron ore and puddling furnace cinder, or slag where old slag banks were discovered that dated from the period of the direct reduction process (bloomery). The pig iron made with ore and slag was also known as 'cinder pig'.

Cupola
This is a shaft furnace with a number of tuyeres around its base and a tap hole for releasing the molten iron. A coke fire was made in the cupola and the charge was pig and scrap iron. Air was blown through to raise the temperature as the coke burnt and melted the iron. Impurities were collected as a slag.

Direct Reduction Process
Various methods in which iron was produced directly from the ore, instead of the method where pig iron is made and purified, such as in puddling.

Edge Tool
A tool, such as a chisel, that is made with a cutting edge.

Fettle
Process of lining a puddling furnace with reactive iron oxide material.

Finery
A furnace, or hearth, that was charcoal-fired and had an air blast, where cast iron is made malleable.

Forepart
Lower front of the blast furnace that may be open or closed.

Forge
1. A furnace, or hearth, where metals are heated or wrought.
2. A workshop where pig iron was transformed into wrought iron.

Forge Hammer
Any hammer tilt, helve or steam used for shingling the puddled balls.

Forging Hammer
Any hammer, tilt, helve or steam used to work iron up to a finished product through hammering, often adding a particular strength to the product.

Forge Train
The mill used in a forge for rolling puddled blooms at the first rolling into a rough, ragged form, known sometimes as muck bar. This was cut up, piled and reheated.

Founder
One who casts metal.

Foundry
An establishment where metal castings are made.

Galvanise
To coat iron or steel with rust-resisting zinc.

Grain Roll
A type of rolling mill roll with a relatively soft surface used for specific types of rolling such as rough work.

Grey Iron
Cast iron where most of the carbon is present as small graphite flakes. This type of iron is the most common found in castings.

Guide Mill
A specific type of rolling mill used for rolling section iron where a box guide is fitted to the mill in order to prevent the metal from twisting as it passes through the rolls.

Helve
A heavy cast-iron hammer, pivoted at one end, worked by cams and allowed to fall by gravity. This was commonly employed in watermills where the turning of the wheel acted on the cams.

Hot Blast
The air that passes through the furnace that has been previously heated.

Hot Blast Stoves
The mechanisms for heating the blast. They were initially iron pipes heated by a burning fuel. Later, these were refractory lined structures.

Ironmaster
A manufacturer of iron.

Ironworks
A building, or establishment, where iron is smelted or where heavy iron products are made.

Lurhrmann Hearth
A special type of enclosed hearth at the forepart of the blast furnace.

Merchant Iron
This term often referred to the iron rolled as sections and sizes such as rounds, hexagons and flats, for example, that would be stocked by an iron merchant.

Merchant Mill
Any rolling mill employed in the general jobbing trade.

R.S. Chattock was an artist who produced a range of prints on local Black Country subjects. In 1872, he published a series of industrial etchings which depicted mining and ironworking in the Dudley area. In this etching Chattock shows a puddled ball of iron being removed from the puddling furnace for transfer to the shingling hammer.

Mill Train
The mill train was two or more stands of rolls coupled together.

Pickling
Treating the surface of the iron with a strong acid to remove scale, rust and dirt as a preparation for further processes such as cold rolling, galvanising and tinning.

Pile
Wrought-iron bars placed one on top of the other ready for reheating and re-rolling.

Plating
A thin layer, or coating, of metal over another metal, or object.

Pig Iron
The product of the blast furnace, when cast in a pig bed.

Puddler
One of two men employed to work the puddling furnace. One was in charge, the other, his assistant, was known as the 'underhand puddler'.

Puddling
This was the purification of impure metal, especially with reference to iron, by agitation of a molten bath of the metal in an oxidising atmosphere. There were two types, dry puddling and wet puddling. Dry puddling was essentially a two-stage process that employed first the refinery and then the puddling furnace. Wet puddling was a single-stage process and involved the use of furnaces lined with a reactive iron oxide (fettling).

Puddling Furnace
This was a reverberatory furnace where broken pig iron was heated and stirred in an oxidising atmosphere.

Puddling Machine
A term applied to a variety of devices employed to mechanise the puddling processs; none were particularly successful.

Purple Ore
Purple ore, also known as 'Blue Billy', was the iron oxide residue left from treatment of pyrites in the manufacture of sulphuric acid. It was used as fettling for puddling furnaces.

Refinery
The refinery, or running out fire, was a hearth in which iron was melted to reduce carbon content and prepare it for dry puddling.

Reverberatory Furnace
A furnace where the fuel and metal are separated from each other and only the heat is made to play on the metal. Flames from the burning fuel are drawn over the metal by a chimney draught arrangement that can be shut off as required by a damper. Air, mill and puddling furnaces all used this arrangement.

Rod Mill
A rolling mill specially designed for rolling rods.

Rolling
The method applied to malleable iron to convert blooms or ingots into bars, hoops, rails and sheets, etc.

Rolling Mill
1. A factory in which metal is rolled into sheets, bars and other forms.
2. A machine used for rolling metals.

R.S. Chattock etching entitled 'Tapping the Blast Furnace', which depicts the flow of iron from the blast furnace to the pig beds.

Coke was produced for the blast furnace in either coke ovens or open coke hearths. This 1872 R.S. Chattock etching of a coke hearth is a reflection of the fact that these hearths were the preferred choice in South Staffordshire.

Shingler

A workman at a shingling hammer.

Stand of Rolls

A pair of rolling mill housings complete with rolls, pinions, etc., through which the metal was passed. It was common to have a number of these connected together as a mill train and the iron was passed from one to the other for shaping, re-rolling, etc.

Steam Hammer

A mechanical hammer powered by steam that was used either for shingling or forging, that was largely superseded by hydraulic or other presses.

Tilt Hammer

A mechanical hammer for delivering light rapid blows.

Tinned

To coat with tin.

Tinplate

A thin sheet of iron or steel coated with tin.

Tuyere

The end of the blast pipe used for conveying air into a furnace. Several were employed around the base of the furnace and were detachable for easy replacement. The increased temperature of the furnace, when hot blast was applied, required tuyeres to be cooled with water.

White Iron

A type of cast iron, low in carbon and silicon, commonly the product of the refinery.

Wrought Iron

Iron that has been made malleable either directly by the finery and chafery or indirectly through puddling.

CHAPTER 1

IRONMONGERS AND THE BAR IRON TRADE

Iron is a greyish metal which, for many centuries, has been produced for domestic and military purposes. Iron ore was found in several parts of the country and the seams that were exposed at the surface were dug out to get the ore. Seams of ore, locally known as ironstone, existed throughout South Staffordshire. The seams which outcropped near the surface encouraged extraction and smelting of the ore from at least medieval times.

The metal was either used for casting or working up to bar iron. As bar iron, it was a useful product for sale and the province of the ironmonger who dealt in metal, selling it onto smiths and other manufacturers.

Iron Production

Extraction of metallic iron was initially a rudimentary process that required charcoal, limestone and the ore.

Three stages were required to convert iron ore into a finished product. The first stage was in the furnace where ore was smelted. Heating iron ore with charcoal enabled the extraction of iron metal. The process was assisted using foot-powered bellows to force air into the fire and raise the temperature. The bellows provided an improved draught, or blast. Iron was produced in an impure form known as a bloom which had to be beaten and hammered to expel slag caught up inside; it was an extremely inefficient method that produced only a few pounds of iron at a time. Bloomeries continued to be the standard means of iron production up to the middle of the sixteenth century. Early bloomeries tended to be itinerate, moving from place to place as ore and fuel was exploited. The introduction of a water-driven bellows created permanent sites for ironmaking.

The existences of several fourteenth-century bloomeries around Walsall, at Bescot and Bloxwich, have been identified from written records. Evidence of a fifteenth-century bloomery was found at Bourne Pool near Aldridge, which belonged to the estate of the Lord of the Manor, Simon Monfort. Archaeologists made a study of the site during 1964, where the slag heap and dam were examined. Document research found evidence that an iron mill at Bourne Pool was operated sometime after 1474, but before 1495 when Simon Montfort was executed for treason. Analysis of the slag suggested iron was made on site, even though the bloomery hearth could not be located, and it was presumed that it had been buried under the lands of nearby Bourne Farm. Other contemporary bloomeries are known to include Nuns Well on Cannock Chase and Coalpool, near Walsall. A bloomsmith would work the blooms on a hearth, passing it on to the brandsmith to complete the process, through hammering it on an anvil into bar ready for the smithy.

The development of the blast furnace in the sixteenth century brought about important changes in iron manufacture when the bloom was replaced by the 'sow' and the 'pig', whilst the higher temperatures made the production of cast iron possible. Blast furnaces came to replace the bloomery, but the transition was slow – even in the early seventeenth century some bloomeries using a water blast were still at work.

Blast furnaces were frequently built near rivers or pools where the power of flowing water could be harnessed by the waterwheel to work bellows and provide a mechanical draught. There were also cases where horses may have been used to turn wheels and activate bellows where water was in short supply.

Each furnace was made in essentially the same basic square design. The raw ingredients – ironstone, charcoal and a limestone flux – were fed into the furnace from the top. The higher temperatures were maintained by air blown into the furnace through tubes known as *tuyères*. Molten iron settled to the bottom where it was tapped off from time to time. Impurities separated from the iron were drawn off separately as slag. It was a continuous process and furnaces were often kept in blast for several months. The amount of iron produced remained relatively small, but the process was a vast improvement on the bloomery. Furnaces were often located near wooded areas where the charcoal was made by a controlled combustion of timber. The ore was brought to the furnace by whatever means of transport was available.

Many accounts of the West Midlands iron trade emphasise the Industrial Revolution and the part the canals and the steam engine played in the growth of ironmaking in the region. There was, however, another 'revolution' – the development of the watermill, which was just as important for it established ironmaking and ironworking in this district. The engineering skills that established the watermills, mill ponds, water channels and the machinery for capturing water power to raise a forge hammer or work the bellows for a blast were essential to ironmaking and working. These skills and tools were available in medieval times when charcoal from the forest provided the carbon needed for smelting iron ore and working that metal up into the saleable commodity known as bar iron.

The second stage in ironmaking was in the forge where pig iron was reheated with charcoal and hammered into bars by tilt hammers operated by waterwheels. This process needed a greater water supply or a faster-flowing stream to generate sufficient power for the work. Two separate departments existed known as the finery and the chafery. The finery took the pig iron to work on. Essentially it was a hearth with an air blast used to make wrought iron through removing carbon from the iron to make a 'bloom'. The chafery finished the treatment. This was another hearth that was fired on charcoal, coal or coke for reheating wrought-iron blooms ready for reworking. The hammer was raised and lowered through the actions of cams on a revolving shaft, each blow coming down on a block that supported the metal being worked. A pair of stout wooden cross timbers acted as the base. Archaeological studies at forge sites in the Weald have demonstrated that considerable damage was done to such timbers through the action of the hammer.

The grade of iron made varied widely so that where high levels of carbon still existed in the metal the piece being worked often crumbled under the hammer and had to be returned to the finery for further treatment. It was then heated in a chafery and repeatedly hammered to expel as much of the impurities (slag) as possible. The stamped iron was next broken into short pieces, drawn out, doubled and welded under the hammer to make the product known as 'bar iron', which was the item that was sold on. Different grades of 'bar' existed and were known as common, best, best best or treble best according to how many times the operation was carried out. Treble best was the highest grade that was achieved through four reworkings of the metal. About 32cwt of pig and about 24cwt of charcoal were generally used to make a ton of bar.

Finally there was the third stage at the smithy where bar iron was heated by coal, as opposed to charcoal, and fashioned into utensils, tools and weapons. The smithy could be anywhere; the forge needed to be within range of small supplies of charcoal, but was usually restricted to a substantial stream because of the power required to raise the heavy tilt hammers; for furnaces localising factors were proximity to ironstone, an abundant supply of charcoal and a small stream to work the bellows.

The first type of ironworks was therefore the forge. In the Middle Ages output of iron was limited; a forge could not make more than 2 tons of bar a week and, with seasonal limitations of water supply to turn the wheel, often only made between 50 to 100 tons a year.

The First South Staffordshire Ironworks

Any comprehensive study of the early ironworks of the region would easily fill a book on its own account. It is the purpose of this author to select certain examples to explain the development that led to the coke-smelting process, which formed the basis of ironmaking here from the 1750s.

Iron working was initially conducted in the vicinity of seams of ore that outcropped or came close to the surface. Such places were to be found in Furness, South Staffordshire and the Weald in Sussex. Much research has been conducted into the history of the Wealden iron industry where bands of ore were dug from shallow pits. Bloomeries had existed in this district from Neolithic times and there was much activity during the Roman occupation where a number of sites have been identified. In the Anglo Saxon period iron working seems to have receded and only one site has been so far attributed to this period. There was a resurgence in ironmaking during medieval times reaching a peak during the sixteenth century. It is now believed that the first British blast furnace was tried in the Weald. The technology had been developed in Europe and came to Britain with French ironworkers who migrated to Sussex.

Within the West Midlands, iron working may have descended from Roman times and some evidence of a Roman cinder mound was discovered near the parish church in West Bromwich during the nineteenth century.

Historians have suggested various ways in which the iron industry has spread in this area. Some believe that it grew up around Wednesbury and Walsall at an early date, and was well established during the Middle Ages, when both ore and fuel were in plentiful supply. This trade then spread out in circles of ever-increasing diameter like the ripples caused by casting a stone into a pool. One group was believed to have followed the Tame Valley towards Birmingham and Tamworth; another group used the tributaries of the River Trent that ran through Cannock Chase. Others suggest a more random arrangement where iron masters capitalised on charcoal and water supply.

There are records of ironworking conducted around Rugeley during the thirteenth and fourteenth centuries, but a key development was made on the Chase during the sixteenth century. During 1546, the first Lord Paget, son of a local nailer, received Cannock Moor as a Grant from the Crown. The earliest direct evidence is that Paget estates included three iron mills in 1553.

The first West Midlands blast furnace was established by Lord Paget in 1561 and was located near Hednesford on Cannock Chase. Charcoal was obtained from the timber of Cannock Forest. Paget's furnace was known as Cannock, or Cannock Wood, and was operated by Thomas, Lord Paget until 1583. In that year Lord Paget was suspected of participation in the Throckmorton Plot to put Mary Queen of Scots on the throne. He fled the country and his estates were claimed by the Crown. Exchequer records for this period show that ironstone was chiefly being drawn from the Walsall neighbourhood, but local Cannock ironstone was also used. The charcoal was produced locally. The records of iron deliveries include several prominent Birmingham businessmen such as Smalbrook and Colmore.

Management by the Crown proved less than satisfactory and the Cannock property was leased to Sir Fulke Greville. His lease of the Cannock Wood estate included two iron furnaces, two iron forges, all waters thereto, a large proportion of the trees in Cannock Forest of Cannock, all formerly in possession of Thomas Lord Paget. Following an inquisition of 1595 it was found that Greville had devastated the region and had not taken the time to coppice and generally ensure the regeneration of the forest; he even felled trees marked for the Queen's use. In 1597 a decision was made to grant back the estate to the Paget family, but with the lack of timber, production of iron was handicapped. Thereafter, it seems that only the forges were kept in operation.

Power for the forges and furnaces were derived from the streams that ran into the Trent near Rugeley. Researchers have established the identity of some six furnaces in the Cannock Region, which operated during the sixteenth century. In addition to Paget's original furnace (Cannock Upper Furnace) there was a Cannock Lower Furnace, a furnace at Cannock Wood (Beaudesert) and also Deepmore, Teddesley Hay and Wolseley Furnaces. Deepmore was located near to the

Wyrley Brook that ran into the River Penk, whilst Teddesley Hay was placed to the east of Penkridge near a stream that also fed the Penk. A seventeenth-century furnace operated at Coven possibly from 1635 until about 1690.

Another local ironmaster was Sir Francis Willoughby who came to develop the Middleton district of North Warwickshire in about 1570. The original works appear to have been a version of a hammer mill where bar iron was made for the smithy. In around 1590 a more ambitious project was embarked upon when a furnace and a forge complete with two fineries and a chafery were built. By 1593 ironstone was obtained from local sources near Polesworth and from 'Heyehead', which is Hay Head, near Walsall. In 1595 an agreement between Thomas Meryhurst of the Delves near Wednesbury and Percival Willoughby enabled the sinking of pits to extract ironstone. Lack of funds and fuel led to the ironworks closing about 1600.

There was another sixteenth-century furnace that was known as Perry Furnace on the Tame near Perry Barr in Staffordshire, which has been traced by researchers to have been in existence by 1591. During the early part of the seventeenth century eight new charcoal furnaces were erected in the West Midlands. These were developed along Stour and Tame valleys using stream or rivers to provide power for the blast. There were furnaces at Aston (Hockley Brook), Bromwich (Tame), Compton (Stour), Cradley (Stour), Grange (Smestow), Hales (Stour), Himley (Holbeache Brook) and Rushall (Tame).

Ironmaking was a slow process. Preparation of the materials, pre-heating the furnace and the smelting process itself all took time to accomplish. Both the limestone and ore had to be broken up beforehand and the ore was also heated with small charcoals or coal to anneal it before it was despatched to the furnace. Good-quality ironstone, which was low in sulphur and phosphorus, was mined by open work in places where the measures became exposed. Such works included the Delves and the Foxyards near Dudley. The iron was carried by packhorse or cart to the furnace.

The manufacture of charcoal required large amounts of wood and this was a finite renewable source. An important supply of wood was obtained from felled trees or from coppices that were cropped on a cycle of between fourteen and eighteen years. As charcoal was required for both the furnace and the finery careful management of resources was needed. By the seventeenth century consideration was given to the use of pit coal and although various attempts were made to smelt it with the ore, little success was achieved. The use of charcoal continued, despite the limit on iron production.

Dud Dudley, Ironmaster

Those who persevered with finding means to use coal as a substitute for charcoal in the blast furnace included local ironmasters Dud Dudley and John Wood. Dud Dudley, a son of Edward Sutton, Earl of Dudley, and Elizabeth Tomlinson, his mistress, spent much of his adult life associated with the iron trade. According to his own autobiography, Dud was educated at Balliol College, Oxford then went on to manage his father's ironworks.

The Dudley estate was rich in coal, ironstone and limestone. There were also extensive woods on Pensnett Chase (Kingswinford) and around Ashwood, Baggeridge, Dudley, Gornal and Himley. Edward, Lord Dudley, is credited with the establishment of four smelting furnaces on his estate at Cradley, Coneygree Park, Himley and, Gornal Wood. There was probably another at Ettingshall. The estate lacked few powerful streams and the forges were located elsewhere. Cradley Forge was located at the perimeter of his land using water from the Stour. Greens Forge was located on another person's land. It was a former corn mill, which had been leased in 1600 and was powered by water from the Smestow.

John Robinson held a patent to smelt iron with pit coal. Lord Dudley obtained a licence from Robinson in 1619 and then renewed the patent for twenty-one years in his own name. The application for the patent was made during February 1621/1622.

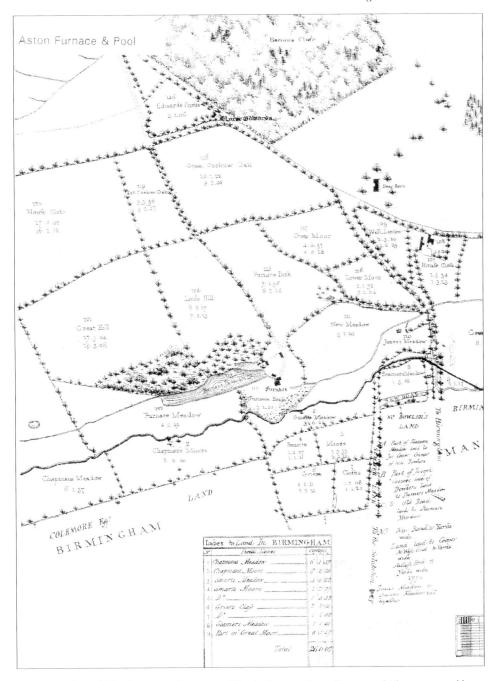

The closest charcoal blast furnace to the centre of Birmingham was Aston Furnace, which was powered by the water from Hockley Brook. The land belonged to the Holte family, whose records are preserved as MS 21 in Birmingham Library Archives. Tenants of the furnace include (1) Humfrey Jennens, (2) Christopher and Riland Vaughton, (3) John Mander and Phelicia Weaman, (4) John Knight and Abraham Spooner and (5) Edward Knight and Abraham Spooner.

Edward, Lord Dudley, treated all his children favourably. Dud Dudley was appointed as manager of the Cradley Furnace and Forge. Much information about the life of Dud Dudley has been gleaned from his own book *Mettallum Martis* published in 1665, which was the basis for the frequently printed facts in nineteenth-century industrial histories that Dudley was the first to smelt iron successfully using coal. During the twentieth century, however, work by A.A. Rollason and R.A. Mott have challenged these views.

R.A. Mott produced a detailed study entitled 'Dud Dudley and the Early Coal Industry' which was presented to the Newcomen Society in November 1934. Mott observed that Dudley's published account was made when he was aged sixty-six and when his records had by his own statement been destroyed during the Civil War.

Dud Dudley's account identifies certain locations that have been proven to exist, although some doubt remains about the accuracy of dates quoted in his account. Dudley said that he came down from Oxford to his father's Pensnett Ironworks in 1619, but Mott checked college records to find that he was first enrolled as a freshman commoner in September 1619 and appears again in the lists for September 1620 and 1621. The date discrepancy is an intriguing one as it opens up the possibility of him being born later than the published date of 1599.

He appears to have specifically been associated with the Cradley Furnace and ironwork, as he mentions the location of his father's works as in Worcestershire. In his words:

> The first patent being granted by King James for 31 years in the 19th year of his reign upon just and true information that the author had the year before made many tuns of iron with pitcole at a furnace or ironworks, in the Chase of Pensnett, in the county of Worcester, beside cast iron works of sundry sorts; and also two forges or iron mills, called Cradley Forges, fined the said iron into merchantable good bar iron.

Mott records the furnace was placed north of Cradley Heath and south of Dudley Wood alongside a stream that fed Cradley Pool. The two forges were identified by Mott as Cradley Forge and Lodge Forge on the River Stour. The relative close proximity of these works would appear to be a sensible working arrangement. It was at the furnaces and works that Dudley made annually 'his great store of iron, good and merchantable together with brewing pots, cisterns and pots'.

Two other furnaces, Coneygree and Gornal Wood, are not mentioned and it can be interpreted that they were not at work. Another possibility is that they were the responsibility of separate iron masters or members of the greater Dudley family.

Mettallum Martis recounts Cradley being destroyed by a flood. This has been traced to Mayday 1623. Dud then moved to Himley Furnace, in Staffordshire. By his own account Dudley fared badly here for the want of a forge to make bars and found great difficulty in selling on his produce. Undeterred, Dud went on to erect another, larger furnace at 'Hasco Bridge'. The location of this furnace has been said to be at Askew Bridge near Sedgeley and further upstream from Himley Furnace. This was a district where he discovered new coal and iron mines. Near here was Upper & Lower Gornal where important seams of ironstone existed.

It was at Hasco Bridge that Dudley was evicted from his furnace by 'riotous persons' and his bellows cut to pieces. Dud Dudley suggests that this was done by rival charcoal ironmasters, but it is recognised now that the instigator was Dud's own father, Edward, exercising his legal rights. By 1622 Edward was in debt and in order to keep the Manor of Himley from his creditors he had made certain grants to Thomas Dudley and William Cox, who later requested Dud Dudley leave the property.

Edward Sutton had five legitimate children by his wife, Theodotia, but only one son, Ferdinando. By his mistress, Elizabeth, he had another four sons and seven daughters. When Elizabeth died in 1629 provision was made from her estate for all her living children except Dud.

The increased access to public records such as Exchequer Papers have in recent times greatly increased the knowledge of the iron trade during this period. Though further work needs to be

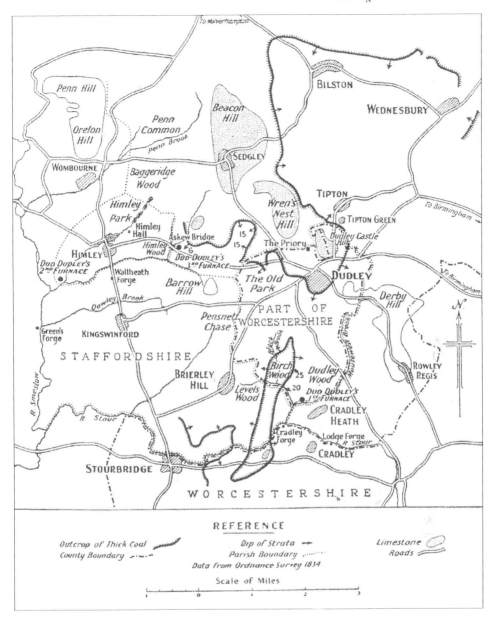

R.A. Mott map of the location of Dud Dudley's furnaces. (Map source: Proceeding of Newcomen Society, November 1934)

done to establish a better understanding of the timeline regarding Dud's various iron ventures. Despite the eviction Dud Dudley continued with the business and fought his father through the courts. A bill of complaint against his father Edward Sutton, Thomas Dudley (his brother-in-law and executor of Elizabeth Tomlinson's will) and Henry Jevons (servant to Elizabeth Tomlinson and also executor of her will) was made in May 1631 where he claimed lawful entitlement to lands, ironworks, iron ore and coal mines in Tipton, Sedgley, Kingswinsford, Rowley Regis and Oldbury. The substance of this complaint also contained allegations that have a more detailed interpretation to those implied in *Mettallum Martis*.

The long-term dispute between Dud and his father Lord Dudley came to a head following his mother's death. Dud Dudley felt that the executors of Elizabeth Tomlinson's estate had not proved the will but had taken possession of the estate. Furthermore Lord Dudley and the executors had attempted to ruin Dud. Allegations included theft of his store of iron, stones and coal from his home in Tipton and then the subsequent possession of his lands and expulsion from the property. The eviction took place in February 1631.

Dud Dudley stumbled from one crisis to another as he pursued his trials with ironmaking. Litigation and debt were part and parcel of an apparently troubled life. Yet such were his skills and knowledge that Dudley forged new partnerships to finance his ventures. In 1638 he was granted another patent:

Patent 117
Dudd Dudley, David Ramsey (a courtier), Sir George Horsey (a Member of Parliament) and Roger Foulke (a counsellor of the Temple) 2nd May 1638

1. Making cast iron and bar iron with seal coal, or pit coal, peat or turf.
2. Refining metals.
3. Also mining and working ores of gold, silver, copper, or lead mixed with silver or quick silver.
4. Draining mines and works for said purposes.

If there was any chance for Dud Dudley inheriting the estate on his father's death, this possibility was lost when these estates were passed over to William Ward, whose son, Humble Ward, in 1628 married Sutton's grand-daughter, the only child of Ferdinando, his heir. Burke's account in his 'Peerage' compilations attributed the marriage to Edward Sutton's need for money. As Ward came from a rich family of London goldsmiths, it was an arrangement that suited both parties. For Sutton it provided a solution to his cash problems and for the Wards it brought an improved position in society with lands and the prospect of a title. For Dud Dudley it was opportunity lost.

Surviving leases and deeds held by Dudley Archives include several properties passing to Ward from the early 1620s. Much of Dud Dudley's property was returned to a William Ward and the Earl of Dudley's estate in a notice to quit dated 11 March 1641 following a ruling of the Court of Chancery. This included mines, furnaces, houses and land in Cradley, the borough and foreign of Dudley; the manor of Rowley; Knoll Pitts (coal mines) at Knoll Hill, Pensnett Chase; coal mines called Disselding, Pensnett Chase; Manor of Himley; Himley Park; Chasboll Chase and meadows in Sedgley; the Manor of Swynford Regis (Kingswinford); Ashwood Chase or Hay; and Baggeridge Chase.

Despite the various arguments made above, a core element of what Dud Dudley wrote about has been seemingly ignored. This was the preservation of charcoal and wood for shipbuilding, particularly in defence of the realm. He maintained that iron made with coal was carried on and worked in four ironworks, Cradley Forge, Greens Forge, Heath Forge and Swin Forge, which had saved 30,000 loads of wood since his invention in 1618. It is a statement that suggests a long-term use of smelting coal with ironstone. Dud Dudley was careful not to specifically mention the method of working. If he still had an interest in the process, it would result in industrial suicide should he mention in accurate detail his process of iron smelting.

Another key element is the nature of the iron made by Dud. Fuel for the furnaces was said to be from slack or small coals of the Staffordshire Ten Yard seam as fuel for blast furnace. Large amounts of slack were made in mining Ten-Yard Coal and the slack at that time had no market. These seams contained sulphur and even a small percentage of sulphur would make the iron 'red short', or 'hot short' which meant it could not be worked when heated in the finery and any attempt to do so would cause the iron to split or crumble. The problem of using coal was solved eventually by coking it. That was burning coals in heaps to drive out the sulphur. This was only possible when the coal is of the 'caking' type. Staffordshire thick coal is 'non-caking'.

It is, however, possible that Dud had a comprehensive understanding of local coals. Not all local coals were non-caking for if that were the case the town gas makers of the nineteenth century would have faced a serious handicap. Yet suitable coals were found locally for carbonisation before transport improved and gas coals could be delivered from further a field. *Mettallum Martis* identifies the different seams of coal and ironstone and indicates that Dud and his contemporaries had a good understanding of the measures. Dud's knowledge of the seams of coal may have been a factor in the search for suitable coals to smelt with coal. Geological surveys have since shown the diverse nature of coal seams that existed throughout the district, and some had reduced sulphur content. There is an element of research that still needs to be carried out and perhaps gain a fresh understanding about what was generally known at the time especially about what was then accepted practice and what was not required to be written down.

The success of coal smelting at this time seems limited and despite the claims for the making of the sought-after bar iron, the iron made by Dud seems to have been principally adapted for the casting of pots and associated ironwork.

The giving up of the lands by Dud Dudley coincided with a move away from ironmaking to that of soldier. Dud Dudley served in the Bishops War against the Scots and then in the Royalist Army throughout the Civil War rising to the rank of colonel. He was captured by Andrew Yarranton, then a parliamentary forces captain, after a skirmish in Boscobel Woods. He was sentenced to death for treason, but escaped and made his way to Bristol where he masqueraded as a Dr Hunt. During his military career he seems to have had little association with the iron trade. Whilst still a fugitive, in the 1650s he attempted to smelt lead in a bell house near Bristol and much later with the restoration of the monarchy he was associated with another smelting venture near Queens Cross, Dudley. This was a horse-mill-powered blast furnace that made iron with 'charcoal made of wood and pit coal'. Dud Dudley's interest in this affair had ended by 1674 when John Finch and Sir Clement Clerke were partners in the enterprise.

John Finch was a member of a family of Dudley ironmongers, who possessed a number of forges. He leased Stourton Mill at Kinver and converted it into a forge during 1670. Finch also took over Cookley Mill during the 1670s. Cookley and Stourton probably became the finery forges for the iron supplied from Queens Cross.

A lease was granted for Coneygree Furnace in February 1631 on the east side of Dudley Castle. This lease was granted by Lord Dudley to William Ward of London for a period of fifty years and William Ward paid a premium of £400 to Lord Dudley; the rental was nominal. The land in the lease was the 'Paddock', called 'Coyneygre', some 30 acres in extent and forming part of the 'parke or ground' on the east side of Dudley Castle, in which a furnace to make iron was erected; with this went mines of coal and iron in the 'Coyneygre' and right was given to William Ward to enter with workmen, horses, carts and carriages to get the coal and iron. An interim memorandum is endorsed on the lease recording an agreement for Lord Dudley to take any coal for the necessary use by himself or his family for firing only in his lordship's house or houses. He was to pay Ward for the getting and digging of the coal, according to the usual rate. As for the furnace, it was to remain 'as it nowe dothe'.

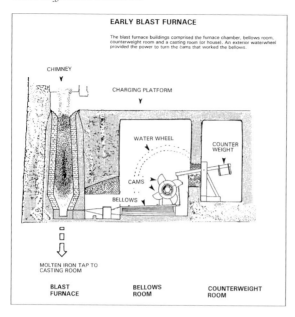

EARLY BLAST FURNACE

The blast furnace buildings comprised the furnace chamber, bellows room, counterweight room and a casting room (or house). An exterior waterwheel provided the power to turn the cams that worked the bellows.

CHIMNEY

CHARGING PLATFORM

WATER WHEEL

COUNTER WEIGHT

CAMS

BELLOWS

MOLTEN IRON TAP TO CASTING ROOM

BLAST FURNACE | BELLOWS ROOM | COUNTERWEIGHT ROOM

A drawing of an early blast furnace, which is based on the surviving Dyfi Furnace, near Machynlleth.

The Charcoal Ironmasters

The period following the Civil War saw rise in the number of charcoal furnaces at work and prosperity for many of the ironmaster families responsible for them. Most notable were the Foley family who worked several local furnaces.

Grange was located on the Smestow Brook and apparently supplied iron to the nearby Heath Forge on the Wom Brook. The early of history of Grange Furnace remains vague, owners of the estate includes the Wrottesley family. They were landed gentry and it was common at this time for the aristocracy and gentry to lease revenue earning properties to professional ironmasters instead of working the property with a manager. It has been suggested that Richard Parkes operated Grange along with his other properties that included the Old Forge, West Bromwich and that Grange passed with the other ironworks to Thomas Nye and then subsequently became part of the greater partnership of Thomas Nye and Richard Foley.

The Coke-Smelting Process

During 1693 'The Company for making iron with pit coal' was formed to exploit a patent granted to Thomas Addison in 1692. They are believed to have owned a furnace at Cleator in Cumberland built in 1694 by Thomas Addison and Richard Patrickson. There was also an association between Addison and Shadrach Fox of Coalbrookdale Furnace who is known to have supplied grenade shells and shot to Bristol. Coalbrookdale Furnace was damaged during 1703 and remained out of use until Abraham Darby recommenced production there. Some historians have suggested a possible early use of coal or coke at Coalbrookdale during Fox's occupation of this furnace.

Meanwhile the search for a successful method of smelting coal with iron ore continued. It was left to the Ironmasters of Coalbrookdale to finally develop a procedure that worked. It is widely accepted that the method that used coke to smelt ironstone was first applied there in 1709.

This method was developed by Quaker Abraham Darby (1676–1717) who was born in Sedgeley. His family had combined the making of nails with farming. As young man he served an apprenticeship with Jonathan Freeth, a maker of malt mills in Birmingham, and had then moved on to Bristol where he set up a brass and ironworks in partnership with others. Darby developed a means of casting iron-bellied pots in sand, which was granted a patent in 1707. This led to a new partnership in Bristol and the subsequent acquisition of the disused iron furnace at Coalbrookdale in 1708.

Examination of Darby accounts in from 1709 has demonstrated that Darby made successful attempts to smelt iron with coke. Researchers found records of coal supplied by Richard Hartshorn & Co. of Little Wenlock. The lack of evidence for coal resale and the specific entries to the payment for coking the coals as well as the lack of entries for charcoal purchase, has led to the interpretation for setting the date of successful smelting of pit coal with ore as from 1709. Coke smelting seems to have been generally confined to the Darby family and their circle of (chiefly of Quaker) friends. The quality and standard of iron produced was variable and it would be some time before charcoal would be superseded in the general production of pig iron.

Coke was produced by burning coal in special hearths. Usually these hearths were located near the furnaces. New furnaces tended to be constructed beside or on the coalfield but still relied on waterpower to create the blast. Charcoal furnaces continued to make iron throughout the eighteenth century in locations where careful management such as coppicing preserved suitable supplies for the charcoal burners. The Dyfi (or Dovey) Furnace, near Machynlleth, is an example, which still survives. This charcoal furnace was in operation between 1755 and 1814. Iron ore was brought by sea from mines near Barrow-in-Furness. Charcoal supply continued to be a limiting factor. The supply was restricted and it has been suggested that make of bar iron throughout Britain was between 250-350 tons a week.

Few coal-smelting furnaces actually existed before 1750 and all appear to have been associated with the casting trade and not the bar iron trade. Apart from Coalbrookdale, there was a furnace at Bersham operated by Charles Lloyd and Richard Wood that smelted with coal from 1722 until the time when Charles Lloyd was made bankrupt and Bersham Furnace reverted back to charcoal. Other coke furnaces operated at Redwood in the Forest of Dean and Little Clifton in Cumberland.

Historians have long speculated on the reasons why coke smelting was carried on such a limited scale. The technical factors controlling the smelting of ore for wrought-iron manufacture were essentially:

1. Internal shape of the furnace.
2. Burdening (that is the proportion of raw materials to each other and coke charge).
3. Method of blowing.

If Abraham Darby had arranged these factors correctly, he should have smelted an iron low in sulphur and phosphorus, which was the type of iron essential for wrought-iron making. The main point of concern was the blast, for coke smelting needed a stronger blast than charcoal smelting. Abraham Darby is known to have improved the method of providing the blast. Those who tried to emulate Darby were less successful. Another crucial factor is that Darby was content to use the iron for casting purposes and had a ready-made product to sell on and did not develop the invention further. Coke-smelting methods for bar iron manufacture continued to challenge ironmasters. Wednesbury ironmaster William Wood took on the challenge and with son John made unsuccessful attempts at coke smelting at Frizington, in Cumberland.

Iron founding began to develop into an industry independent to wrought iron manufacture and the founder developed techniques as an alternative to wrought iron. The inventive were granted patents in this line of manufacture. These included Richard Baddeley, ironfounder and button maker of Birmingham and Rushall Furnace who in May 1722 patented a new method making 'streaks' that is the iron rim of a cart or wagon wheel. The patent also included a new type of box smoothing irons be made of metal will that 'will be so compact a body as to retain the heat much longer than those made of barr iron and thereby become more useful and beneficial'.

Bar Iron Manufacture

Meanwhile the trade in making charcoal bar iron were developed along profitable lines by the Foley, Knight and Lloyd families using water-powered ironworks throughout the Midlands.

One of the basic uses of bar iron was its slitting into nail rods for making nails. The means of cutting the rods was hard won. The slitting mill was an important development for the iron masters. It was, perhaps, with the exception of the mechanical hammer, the first piece of ironworking technology to be introduced. More importantly it included elements of the rolling mill, which was to become a fundamental tool of industry.

The slitting mill consisted of a pair of iron shafts mounted vertically above each other, each carrying a number of sharp-edged discs. The shafts were connected to the waterwheel that turned them in opposite directions to each other and by this means the discs acted as a rotating cutter that sliced through the piece of iron that passed between them. As the shafts rotated the iron strip would be drawn in and slit lengthwise into strips. The strip width being dictated by distance by which discs were separated on the shafts. In line with the cutter discs and driven by the same waterwheel were two plain cast-iron rolls, also rotating in opposite directions. A piece of iron passed between these rolls would form a wide flat strip suitable for slitting.

The first British slitting mill became operational at Dartford, Kent in 1590, and first in the Midlands includes a slitting mill at Rugeley on a tributary of the Trent and a mill at the Hyde at Kinver on the River Stour. Richard Foley established Hyde Mill in 1628. A very colourful account of how Foley acquired the technique is now treated as a work of fiction. Griffith's *Guide to the Iron Trade* (1872) has Foley venturing to Russia to steal the secret in disguise as teacher of French and playing the violin to supplement his earnings whilst away from home.

Slitting mill technology for rolling bars before slitting was later adapted for rolling thin sheets that formed the basis of a new industry known as tinplate. Thin sheets of iron were covered with a layer of tin to protect it from rust. Tinned iron was a German invention and had originally been applied metal shaped under the hammer. It was at John Hanbury's ironworks at Pontypool during 1690s that the first British tinplate was produced. Here was perfected the use of a water-driven rolling mill to produce thin iron sheets that were used to make pots, pan, kettles and also tinplate.

In 1627 Hyde Mill was a fulling mill on the Stour. Within a year Richard Foley of Dudley had taken possession and rebuilt the mill as a slitting mill. About this time Foley moved to Stourbridge. Later 1630s Hyde Mill was worked by Foley's brother-in-law George Brindley who lived in nearby Hyde House. George's son Richard bought the mill in 1647. In 1651 Richard Brindley agreed to slit 8 tons of iron a week supplied by Thomas Foley up to a total of 300 per year. In 1653 and 1654 the mill also slit iron from Sir Thomas Middleton's Shropshire forges.

The Brindleys continued to work it until 1731 when Richard Brindley's grandson was declared bankrupt. He died the same year. In 1733 the mill was bought by George Draper, whose son George sold it in 1741 to Jeremiah Caswell, the tenant since 1736. Caswell died in 1769 and daughter Eleanor ran it at least until her marriage in 1776 to Paul White Curate of Kinver. She was running it in 1780 the year after husband's death and in 1781 she went into partnership with her nephew Francis Homfray of Gotherseley.

Richard Foley started to add other properties to his industrial empire including a charcoal blast furnace probably worked by Dud Dudley. The agreement was made on 30 June 1628 between Edward Lord Dudley and Richard Foley of the furnace in Himley Park for ten years.

Principal members of the Foley family included Richard Foley (1588–1657), founder of the family fortune who held numerous ironworks including Hyde Mill, and Thomas Foley (1617–1677), Richard's son, who extended the network of Foley ironworks and also leased Cradley Furnace from 1662. Other family members included Robert Foley (1627–1677), brother of Thomas and ironmonger of Stourbridge, Paul Foley (1650–1699) and Philip Foley (1653–1716), brother of Paul and an ironmaster of Prestwood, Staffordshire.

Foley money controlled two important partnerships in the Midlands:

1. Ironworks in Partnership: Four furnaces, thirteen forges, four slitting mills and a warehouse. These ironworks were situated in Forest of Dean, the Stour Valley and as far afield as Pembrokeshire
2. The Staffordshire works: Controlled Mearheath Furnace and forges at Oakamoor, Consall, Chartley, Bromley and Cannock and slitting mills at Rugeley and Consall

The 'Ironworks in Partnership' was set up in 1692 and comprised a partnership of Paul Foley, Philip Foley, John Wheeler, Richard Avenant and Richard Wheeler. John Wheeler was appointed cash holder at a salary of £200 per year. He resided at Wollaston Hall near Stourbridge.

West Midland Ironworks owned by the Foley Partnership 1692
As per BLC Johnson

Furnaces	Hales & Grange
Forges	Cookley, Cradley, Stouton, Whittington, Wildon & Wolverley
Slitting Mills	Cookley, Stourton, Wildon & Wolverley
Warehouses	Bewdley

Philip Foley operated charcoal furnaces at Grange and Hales and supplied iron to local forges within the 'Ironworks in Partnership' group. Little Aston Forge was one such recipient of the iron. Placed on the same brook that fed Bourne Pool, Aston Forge was placed north-east of Aldridge and drew water from Hammer Pool. The land was acquired by Thomas Fowke in 1574 and the establishment of the chafery, forge and hammer mill is believed to be after that date. By 1600 the forge was in a derelict state but was restored by Thomas Parkes who took the lease. Subsequent owners were his son Richard, Richard Middletons and then the Foleys. They made bar iron at Little Aston and much of this went to Bustleholme Mill.

Richard Knight was a partner in the 'Ironworks in Partnership' at a later time. He was born in 1659, the son of Richard Knight of Madeley, Shropshire. Richard's experience in iron trade was to work at the Lower Coalbrookdale Forge. He then went onto take over Morton Forge in Shropshire, where he met and married Elizabeth Payne, daughter of Andrew Payne, ironmaster of Shawbury.

Between 1695 and about 1710, Knight operated Flaxley Furnace in Forest of Dean, which had been acquired from the Foley Partnership. He also was in partnership with Thomas Lowbridge at Ruabon Furnace, but this ended in 1696 when Richard Knight gave up his share. The Bringewood ironworks, Shropshire were leased to Richard Knight around 1698 and he also started production at Charlecote Furnace near the Clee Hills in Shropshire. By 1712 Charlecote supplied pig iron to the Knight family forges at Bringewood and Morton.

The Foley 'Ironworks in Partnership' by 1710 had different shareholders, which included executors for former partners now deceased: Thomas Foley, Philip Foley, the executors of John Wheeler, executors of Richard Avenant, Richard Knight and William Rea.

Stourton Mills were leased in 1670 by Wortley Whorwood to John Finch of Dudley with liberty to convert them into a furnace or forge. By 1671 Finch had a forge there. Philip Foley granted a new lease in 1672 but Finch surrendered it in 1673. The continual action of the tilt hammer threatened foundations of Stourton Castle. The forge then had various owners until the 1680s when John Wheeler and Richard Avenant held it. From 1692, Stourton Forge became part of the 'Ironworks in Partnership'. A slitting mill was added in 1698, when Richard Wheeler held the lease. From 1703 John Cook held the lease and the ownership remained with the Cook family until 1762. Thomas Hodgett had the mill until 1781 and from 1792 it was in the hands of the Homfray family, who turned the watermill into a rolling mill.

Thomas Foley and later Richard Foley were owners of Bustleholme Mill. In 1691 the premises was described as having both a corn mill and slitting mill. During the early 1800s these mills were purchased by James Smith of Hall Green House, who leased them to Thomas and William Morris, ironmaster of Bradley, in around 1819. The Morris's converted Bustlehome Mill into a rolling mill.

By 1725 Knight had left partnership to concentrate on his own iron working ventures. The 'Ironworks in Partnership' also started to give up its property to other ironmasters. During the 1720s Knight was a principal member of Willey Ironworks partnership near Broseley, Shropshire. In 1733 the shares in Willey were distributed between Richard Baldwin, Thomas Green, Richard Knight, Edward Baugh and Mr Payne.

Richard Knight and son, Edward, became partners with owners of Hales furnace, which had been operated by Thomas Littleton & Co. The Knight family formed a new company known as the Stour Partnership and brought with them forges at Cookley and Whittington which guaranteed a market for pig iron produced at Hales. Richard Knight entered into this partnership sometime before 1726 and on Lady Day in 1727 the partnership consisted of Sir Thomas Littleton of Hagley Hall, Joseph Cox, the executors of Clement Acton and the Knights.

Edward Knight, in the late 1730s, built a rolling mill at Bringewood to enter the tinplate trade – the production of tin plate guaranteed a market for large amounts of bar iron produced at Bringewood. The rolling mill produced sheet iron (known as blackplate), which was transported to Mitton Lower Forge on the Stour for tinning.

The Cookley ironworks were founded towards the end of the seventeenth century, and from that time Cookley became a centre of the iron and tinplate industry. Joseph Piper, a native of the village, invented and perfected the patent tinning process. The mill at Cookley was originally a corn mill but had been converted into an iron mill before 1706. These mills were subsequently leased by the Knight family as part of the Stour Partnership. Edward Knight owned the mills from about 1750 through to 1780.

The mills, known as Lords Mill, Wolverley, were corn mills converted into iron mills, and a lease of the forge was granted in 1727 to Edward Knight. In 1669 another iron mill was erected by Joshua Newbrugh and Philip Foley at the Lowe.

In 1619 George Taylor granted a lease of Whittington Mill and converted it into a forge; Richard Foley had taken it over by 1628. During 1640 Richard Foley conveyed his interest in this estate to his son Thomas, who was granted a lease for life by Henry Grey, Lord of Whittington in 1649. The forge was taken over by Thomas's son Philip in 1669 with Thomas Jukes as manager. By 1683 it was held by Richard Avenant, John Wheeler and Andrew Bentley. Between the years 1692 to 1705 Whittington Mill was run by the 'Ironworks in Partnership' and from 1725 operated by the 'Stour Valley Partnership'.

The Attwood family became closely involved with the iron trade from the 1770s when George Attwood and his family settled near Halesowen. George had several sons, the most notable of whom was Matthias Attwood, a man of considerable ability. He was generally reputed to be a hard, stern man, but he was capable of inspiring much affection in many of his dependents. He was a steel manufacturer and also engaged in the nail trade and other industries. In 1791, in conjunction with Isaac Spooner, he established the firm of Attwoods, Spooner & Co., a bank that had a high reputation until failure in 1865. Matthias was the owner of Hawne, the Leasowes and other estates and was a deputy lieutenant or magistrate for the Counties of Shropshire, Staffordshire, Worcestershire and Warwickshire. He took no part in politics but was a decided conservative. His son, Thomas Attwood, became a famous politician serving the town of Birmingham.

A considerable number of forges, rolling mills and slitting mills were set up alongside rivers and streams in South Staffordshire and East Worcestershire during the seventeenth and eighteenth centuries; some were conversions from corn or fulling mills, others were new structures. They all made a contribution to the increasing bar iron and finished iron trade. Ironmasters included Charles Leonard who operated the twelfth-century watermill known as Sandwell Mill. In 1790 Sandwell Mill was described as a mill for rolling and slitting iron and was the property of Charles Leonard. This seems to be the same Mr Leonard who worked Friar Park Forge and Bustleholme Mill during this period.

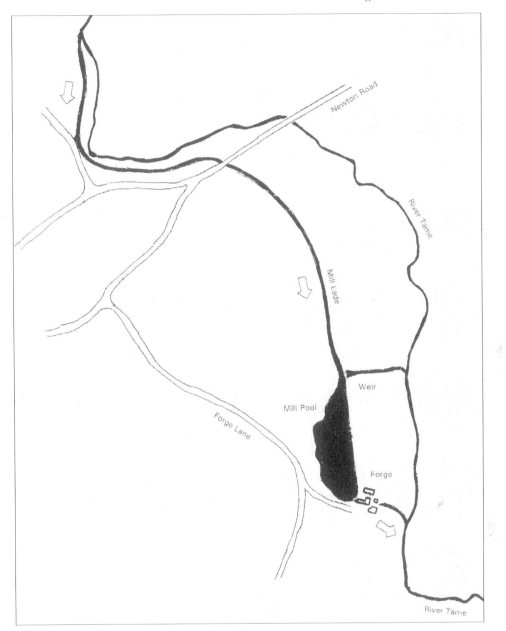

Plan of West Bromwich Forge showing the arrangement of the water supply. The operation of a watermill was dictated by supply. Where a river or stream provided a sustained flow waterwheels might be placed alongside the stream, but in many cases, such as West Bromwich Forge, mill owners had to resort to an extensive arrangement of mill lades, floodgates and a mill pool in order to store enough water for the commercial operation of the mill.

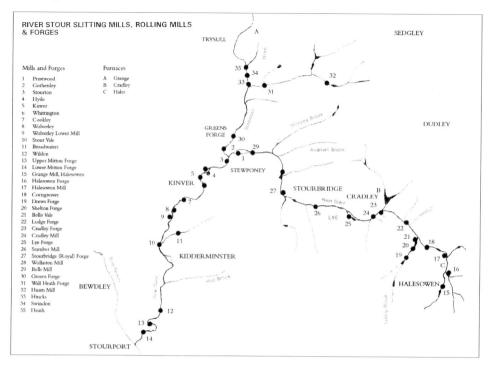

Stour Valley watermills, eighteenth century.

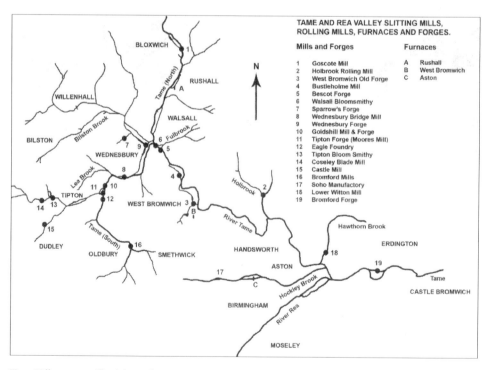

Tame Valley watermills, eighteenth century.

The Old Forge Mill, West Bromwich was adapted by Thomas Parkes who obtained permission from Walter Stanley in 1585 to dig the necessary mill lade and erect a forge, furnace and cottage on the site. After the death of Thomas Parkes the business was continued by his son Richard Parkes and *his* son Thomas Parkes continued trading until 1625. Richard Foley ironmaster leased the forge from Lady Day 1625 and it then remained in occupation of Foley family for a number of years. The Foleys 'Stour Valley' properties at this time including forges at West Bromwich, Wednesbury and Little Aston and a slitting mill at Bustleholme, eventually passed down to Philip Foley of Prestwood. During September 1669 a twenty-one-year lease was granted to Philip Foley that including the forge hammer mill consisting of two fineries and a chafery called West Bromwich Forge.

Subsequent owners of the Old Forge included John and Thomas Adams from March 1725, Thomas Powell, iron master of Dudley briefly, then Edward Kendall who purchased pig iron from Hales Furnace and (probably) bar iron from Stour Valley Forges. In 1742 a lease was arranged with John Churchill of Hints Forge. Wright and Jesson, two local West Bromwich families, then took charge of the forge.

The Jesson family included Thomas Jesson, ironmonger, and son Joseph Jesson. Joseph took up the family business about 1750 and from 1750–60 was buying rod iron from the Stour Valley Forges. During 1762, Joseph took into partnership younger brother Richard Jesson. Ann Jesson, their sister, had married John Wright of the Manwoods Handsworth, from another ironmonger family, and he combined the tasks of nail maker, farmer and steward of the Earl of Dartmouth's estates.

The Parkes family, who played such an important part in development of the iron industry in the district, acquired Willingsworth Hall in 1598. In 1781 the mill on the Leabrook was known as Willingsworth blade mill.

Hockley Brook provided the power for Aston Furnace. It can be traced back to 1615 when a William Cowper occupied the property. Later the Jennens family worked this furnace in connection with Bromford Forge. During 1746 both Aston Furnace and Bromford Forge were added when Knight went into partnership with Isaac Spooner. They arranged for the lease of these properties from the Holt family. Aston Furnace was the closest working blast furnace to the old town of Birmingham. Bromford Forge was placed on the River Tame to the South of Erdington and worked up Aston pig into bar iron.

There were many other watermills placed around the district that were adapted for metal working using the power of the Rea, Smestow, Stour and Tame or their tributaries. In addition to working iron at forges, rolling mills and slitting mills, others were adapted as blade mills that made edge tools and weapons. There were also mills for boring gun barrels, whilst some rolled and shaped brass and other metals to supply the varied 'toy' industries of Birmingham such as buckles, buttons, pins and thimbles.

Advances made in the iron trade were somewhat localised. It would be another fifty years before really satisfactory bars could be produced from pig smelted with coke, though coke had been used to some extent for the chaferies and even charcoal pig was converted in the fineries by coke instead of charcoal. The terms charcoal finish and coke finish are still used to describe the quality of tin plates, though neither charcoal nor coke is used in their production.

CHAPTER 2

SMELTING WITH COAL
THE YEARS OF NEW ENTERPRISE, 1756–1814

Blast furnaces became more common towards the end of the eighteenth century. The increase in their numbers was chiefly due to the perfection of a steam engine that was capable of producing a mechanical draught. It was an invention that enabled furnaces to be built at locations which were not dependent on water power, such as where the ironstone coal was mined, and perhaps more importantly were not dependent on transport routes such as turnpikes or canals.

The road to a satisfactory coke-smelting process proved to be a long and weary one beset with pitfalls. Problems continued with finding suitable coals for caking, and smelting was also handicapped through the means of finding a satisfactory type of bellows for the blast. John Wilkinson (1728–1808) was the first to establish a coke/iron-smelting furnace in South Staffordshire. It was built at Bradley, near Bilston, between 1757 and 1758 and, at first, despatched and received material by road.

This was an area where coal and ironstone was already mined, but there were few streams to supply the water power for the blast. James Brindley's first survey of the Birmingham Canal (1767) shows the location of Wilkinson's iron furnace, which was close to the intended route of the canal and equidistant between Bilston and Wednesbury. If water power was not available the question must be asked as to how the mechanical blast was achieved. During this period Newcomen steam engines were already employed to drain mines in the district; they worked on the reciprocating principal where the beam was raised up and down. The action of the beam could also be used to power bellows, with the result that a reliable and more powerful blast would be available for the furnace. As there is a lack of evidence to support the assumption that a watermill was erected here, the conclusion would be steam power was employed instead, and if this line of reasoning is taken forward it would make Bradley a special place in the development of the iron trade, deserving more recognition than historians have given it.

John Wilkinson, Ironmaster, and the Contribution to Improvements in Blast Engines

Many alterations were made to the original line of the Birmingham Canal before it was completed, and although it passed through the Bradley Colliery Estate the route was at a higher level and went to the south and west of Bilston instead of the east and north as originally suggested by Brindley. Wilkinson's furnace, the Lower Bradley Furnace, was left isolated from the canal for a number of years until a short and private branch of the canal was made near to the furnace yard, in around 1805. Wilkinson built other furnaces closer to the canal and finally three ironmaking plants were in operation (Lower Bradley, Upper Bradley and Hallfields). There was a foundry next to the towpath, where iron goods could be loaded straight into boats. Wilkinson cast many of the parts required for James Watt's steam engines at Bradley and for his other works at Bersham (near Wrexham) and New Willey (near Broseley).

John Wilkinson grew up with the iron trade, learning the methods of casting from his father Isaac Wilkinson. Isaac had been born in Washington, County Durham, in 1695. He belonged to a family of dissenters and came, like other notable ironmasters, to follow the Wesleyan beliefs. Isaac moved to Cumberland to work at Cookson & Co.'s Brigfoot Ironworks, near Workington, where they were smelting iron ore with coke. He was working there during the early 1730s when the coke-smelting method was still being perfected. Isaac became a successful pot founder at Brigfoot and with this experience moved south to Furness in North Lancashire in 1735. Here he worked for the Rawlinson family who owned the Blackbarrow Co. and built up a reputation for casting iron from the charcoal furnaces. During 1748 Isaac went on to manage his own ironworks business, working first from Wilson House in Lancashire and later moving his family to North Wales to continue iron founding and canon boring at Bersham. Isaac was also a founding partner of the Dowlais Furnaces at Merthyr Tydfil in South Wales.

Isaac was a talented inventor and made some important contributions to the trade, such as the patent for improved furnace bellows in 1757. John and his younger brother, William, carried on the trade and became particularly adept at the accurate boring of canons – this skill was translated into the boring of cylinders for steam engines. John Wilkinson had a share of other ventures including the Willey Furnace, New Willey, Snedshill and Hadley Furnaces in Shropshire. Bradley became John's largest and most extensive operation, comprising furnaces and mills.

Details of Isaac Wilkinson's bellows patent was printed in the *London Gazette* (June 1757), the brief description mentions that the bellows could be worked by either water or a fire engine. This statement implies that fire engines, then exclusively of the Newcomen type, were able to work bellows for the blast; this statement lends further support to the theory that the Bradley Furnace had a steam engine for the blast. Historians such as Dr Peter King discount this possibility on the grounds that the stroke of the Newcomen Engine was too slow to provide an effective bellows blast. An alternative suggestion was that pumped water was delivered into a cistern to drive a waterwheel for the blast. At Bradley, the supply of water could have been from the mines. Nevertheless, the theory of direct drive from an engine requires further investigation, as does the required capacity of the cistern to drive a waterwheel effectively.

This study would not only benefit how Bradley Furnace was worked, but also New Willey Furnace, which was also constructed during 1757. A fire engine was erected there, even though there were pools near the furnace and Dean Brook passed through the site. The brook evidently provided power for a forge and boring mill. The furnace, however, was set against the hillside and at a distance from the available water supply. The purpose of the Newcomen Engine has again been suggested to pump water, which in this case came from the pools, into a cistern for a waterwheel. In 1776 New Willey Furnace was the first blast furnace in the country to employ a Boulton & Watt blowing engine. This engine remains the fist accepted example of a direct drive for the blast.

Bradley was only part of the extensive ironworking empire that was operated by John Wilkinson who, like his predecessors the Knights and Foleys, had ironworks and furnaces in different parts of the country. His association with the Birmingham engineers Matthew Bolton and James Watt became a crucial factor in the use of steam engines in ironmaking.

Once the coke-smelting process was perfected, the method was applied across the country. Whether conducted in South Staffordshire, Lancashire, Scotland or Yorkshire, the basic ingredients of coal, ironstone and limestone that formed the recipe for iron manufacture were brought together in the furnace yard. The ironstone and limestone were broken or crushed nearby. Coal was converted into coke at the coke hearth and mixtures of coal and coke were added to the mix that went to the blast furnace. It was also common to heat limestone, ironstone and coal together by a process called 'calcining' before the stone went into the furnace. Within the furnace a chemical reaction was taking place, fuelled by the oxygen in the blast air. The iron compounds were reduced to metallic iron, which collected at the base of the furnace. Above the iron was a slag which comprised all the waste products and impurities left from the stone.

Blast furnaces were worked continuously and tapped every twelve hours. The molten iron was run into specially shaped moulds of sand, which were known as the 'pig beds'. Here the iron

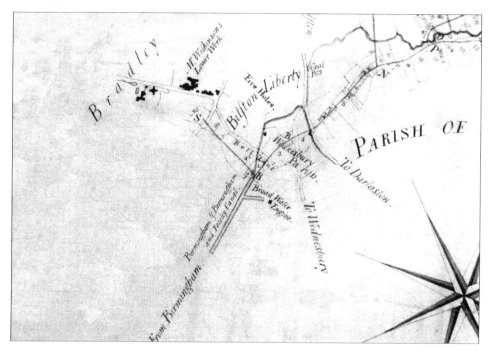

Wilkinson Lower Works as shown on a map for the Walsall Canal. (Source: Deposited plan for Walsall Canal, 1794. Original at Staffordshire Records Office)

was allowed to cool. Pig iron was so called because of the nature in which it was made. The sand beds covered an area of the yard facing the furnaces. Channels were fashioned in the sand that carried the stream of molten iron into long rectangular-shaped moulds. It was part of the furnace labourer's duties to follow the stream and divert the flow whenever the moulds became full. Finally, the whole bed was filled with shapes of iron joined to the original stream of iron in a form that resembled a sow and her suckling piglets, hence the name pig iron. The longitudinal shapes were called the pigs and the connecting, and now solid, stream from the furnace was the sow.

The manufacture of pig iron was only part of the iron-manufacturing process. It could be reheated in a cupola and poured into in moulds. The finished product, in this case, was cast iron. Other iron was passed to the finery and chafery, as had been done in charcoal iron production, to be hammered into wrought iron.

Homfray, Jesson, Knight, Lloyd, Parker and Wood: Eighteenth-Century Ironmasters

Influenced, perhaps, by demand for iron during the American War of Independence (1775–1783), ironmasters looked to South Staffordshire for new sites for erecting coke blast furnaces. Local forges still preferred charcoal iron for most purposes, and charcoal furnaces such as Aston, Grange and Hales remained in work until the 1770s. By this time, however, several coke-fired furnaces were operational in Shropshire and it was a Shropshire family who decided to emulate Wilkinson and build a new furnace at Tipton.

This furnace was built at Tipton Green for George Parker & Co. and appears to have been operational by 1781, when iron was first supplied to the Stour Partnership forges. The Parker brothers had learnt their trade working for the Darby's of Coalbrookdale. George was the eldest, followed by Richard, John, Abraham and Benjamin Parker. The original Tipton Furnace lay close

John Wilkinson's house at Broseley.

to the Birmingham Canal, opposite the junction of what was then a private branch of the canal that served Lord Ward's Tipton Colliery and underground lime works, but was later to be incorporated into the line of the Dudley Canal. A second furnace was erected on this site and there was also an ironworks.

Aston, Grange and Hales furnaces all ceased production during this period. Aston stopped iron smelting between 1772 and 1773 and was converted into paper mill, the water-powered Hales Furnace site was adapted as a forge about the same time. The isolated Grange Furnace was converted for other uses and became known as 'Furnace Mill'. The last people to work Grange Furnace were Thomas and Walter Jorden, who held two shares, with another share owned by Richard Jorden, proprietor of the Heath Forge. Grange Furnace had ceased production by 1772 and the property was advertised for sale in November of that year. The Jordens were reclaiming the cinder bank using the watermill to stamp cinders and reclaim some of the iron trapped there.

Mills on the rivers Rea, Stour and Tame, as well as their tributaries, were sought after for ironworking, competing with other uses for the mill such as the grinding of corn for flour, or the sharpening of edge tools. The Stour Valley Ironworks were particularly busy and many even came to adopt steam power to supplement the waterwheel. At the Hyde, Kinver, the original watermill site was extended to the banks of the Staffordshire & Worcestershire Canal, where a canal basin serving the works was made.

Eleanor White and Francis Homfray operated the Hyde Mill near Kinver. White left the partnership in 1793 leaving Homfray in sole charge. By 1791 Homfray had added a rolling mill and by 1797 had installed a steam hammer. In 1810 the mills were offered for sale and the engine, forge, rolling and slitting mills were included in the sales particulars. Thomas Homfray bought the works. Later, he was declared bankrupt and Hyde Works was offered for sale in October 1819. The next owners were Smith & Shepherd, who also operated the Brierley Ironworks and Colliery, near Dudley. They were bankrupt in 1822. The Stourbridge banking firm of Thomas Hill, Thomas Bate and William Robins then bought Hyde Mill. Stourbridge ironmaster, Thomas Morris, leased the mill and in 1824 decided to dispense with part, if not all, of the waterpower system then employed there. The plant for disposal included an undershot wheel of 22ft diameter and 3ft 9in wide, attached to a driving wheel of 10ft 8in diameter, working onto a spur of 4ft 3in diameter, four arm cams and arms, standards, helve and harness.

The Knight family operated the nearby Whittington Forge. In 1770 Edward Knight secured permission from Lord Stamford to adapt a section of Whittington Forge as a slitting mill. The Knight family worked the mill until 1810 when it reverted to a rolling mill, also used to make wire. During 1753 John Homfray converted a mill on the Broadwaters into an iron forge. These works became known as 'Broadwaters Ironworks', and like Hyde and Whittington, benefited from the canal trade generated by the Staffordshire & Worcester Canal. The Heath Ironworks were located on the Smestow Brook and comprised a water-powered forge and rolling mill. During 1826 the stock and mill was advertised for sale by the assignees of Gibbins, Smith and Goode. Mill and forge machinery included a 18ft diameter by 6ft 8in wide waterwheel linked to 16ft-diameter fly wheel, with iron arms and rim; one 1ft-diameter spur wheel; one 12ft driving wheel; one 7ft-diameter spur wheel; one large iron helve with housing; complete anvil, block and cap and a cam with standards. The rolling mill included grooved, billeting and plane rolls.

Between 1766 and 1767 the Wright and Jesson families decided to take over the West Bromwich Forge and purchased iron from Aston Furnace. The firm was essentially a partnership between John Wright and Richard Jesson. In 1774 they had a Letters Patent for malleable iron from pig iron. John Wright died in the mid-1770s and Richard Wright took over his father's share. Loans were made by Joseph Jesson to maintain the business and promote a new venture at Wren's Nest Forge, Shropshire, on land leased during 1775, and later the pair of blast furnaces, known as 'Barnett's Leasowe', were constructed near Broseley, Shropshire.

By 1798 Richard Jesson was living at the Leverets, in Handsworth. He became High Sheriff of Staffordshire in 1804 at age sixty-three. Elizabeth, a daughter Richard Jesson, married Samuel Dawes, a Birmingham sword cutler. Samuel and his brother, William, had a sword-manufacturing and gun-making business in Snow Hill, Birmingham, and also operated Moor Green Mill on the River Rea as a blade mill.

Barnett's Leasowe Furnaces were placed near Jackfield, near to the south bank of the Severn, which provided a principal means of transport for the iron made at the furnace. Jesson's property at Wren's Nest comprised three forges worked by water and steam power. The forges were located on land between the Broseley to Bridgenorth turnpike and the Severn, and about three miles south of Barnett's Leasowe Furnaces. The head and fall of each mill varied between 20ft and 30ft. The water acted on overshot wheels that in turn operated helves and hammers. A Boulton & Watt engine was used to drive a shingling forge and rolling mill as well as blowing the finery.

Samuel Dawes joined the Jesson & Wright ironworking partnerships of Staffordshire and Shropshire during 1809. The new agreement, dated 12 October 1809, quoted the partners as Richard Jesson of the Leverets, Handsworth, Richard Wright of Manwoods, Handsworth, Thomas Jesson of West Bromwich and Samuel Dawes of Birmingham. Samuel and Elizabeth later made their home at the Leverets, which was a large house with grounds beside Camp Lane.

Barnett's Leasowe Furnace was subsequently leased to Charles Phillips and William Parsons, and the Staffordshire forge business was moved to Bromford Mill and leased to Jesson and Dawes, later S. and J. Dawes. By 1818 West Bromwich Forge was worked by Charles Bache, ironmaster. Bache worked the slitting mill, obtaining bar iron from John Bagnall & Sons of Golds Green. In 1826 Bache was made bankrupt. He was then associated with the metallic bedstead trade, probably at West Bromwich Forge.

The Wednesbury Bridge Mill had been both a corn mill and a fulling mill, but was adapted as a forge when owned by John Wood, son of William Wood (who made the family fortune as a contractor for Irish coinage). Iron was purchased from the Aston and Hales Furnace for use in the forge. John Wood experimented with the ironmaking process and was linked with various patent applications, including that of 1761 (759): making malleable iron from cast iron using pit coal.

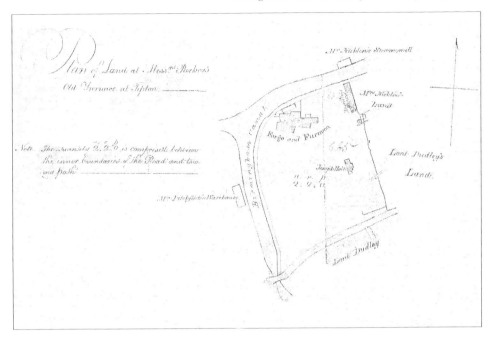

Map of George Parker's furnace, Tipton, 1803.

John Wood was the last tenant of the Aston Forge and following the work of his father, William Wood, presumably used coal at both Little Aston Forge and Wednesbury Mill. He was also a partner with George Attwood at Corngreaves Steelworks. Wood died and was buried in Wednesbury 1779. Little Aston Forge was left standing but was shortly sold on to be converted into a corn mill. Wednesbury Bridge Mill was offered for sale in 1816 and was converted back to a corn mill by Whitmore Jones in 1818.

Wednesbury Forge was constructed at the confluence of the Tame and Willenhall Brook, where a complex of mill pools was formed. There was a lease to Thomas Foley in 1657, then Philip Foley, and other subsequent owners followed. Edward Elwell had this forge from 1817. In 1831 the premises were described as a forge, iron mill and grinding mill

In an inventory of Tole End Mills for Mr Richard Foley senior, a disused watermill and a windmill are mentioned. Snape's Map, 1781, shows a Tipton Forge and Slitting Mill. On this site, or close to it, was the mill known as 'Moore's Mill', belonging to Daniel Moore. In 1798 Stebbing Shaw mentioned one slitting mill worked by water with two undershot wheels to which a steam engine was added. Daniel Moore of Tipton, ironmaster, died in around 1803, but the mill was still known as 'Moore's Mill' in 1812; this name was also given to a brick bridge over the nearby Walsall Canal. Parkes' history of Tipton suggests these mills might also be the location of Taylor's Foundry in Toll End.

A few hundred yards downstream was Golds Mill, or Golds Hill Mill. According to Reeves' *History of West Bromwich* (1836), Golds Hill Ironworks was originally a slitting mill, but was subsequently converted into a forge. There were two mills placed close together, their history is described in more detail later.

The Lloyd family operated charcoal ironworks in different parts of the Midlands and Wales: Kings Mills (Trent), Melbourne (New Brook), Burton Forge (Trent), Town Mill, Birmingham (Rea), Powick (Teme), Mathrafal (Vrynwy), Dolobran (Vyrnwy) and Bersham (Clwedog). As described in 1803, the mills at Powick comprised a forge with three fineries, a chafery, a baling furnace and two hammers. These works were capable of making from 400–500 tons of bar iron per year. The adjacent rolling and slitting mills had a greater capacity; they could process from 2,000–3,000 tons on 'half blooms' per year, slitting rod iron and rolling black plate, hoops and other iron. Such was the flow of the River Teme that these mills were in regular use.

Development of Coke Blast Furnaces in South Staffordshire

Between 1780 and 1800 a select group of coke blast furnaces were established along the old line of the Birmingham, Dudley and Stourbridge canals. These included a furnace at Deepfields, Coseley, for Stokes & Pemberton, and furnaces in Bilston for the partnership of Bickley and Gibbons, which were located beside the canal on the west side of Bilston parish.

George Stokes was working Coseley Furnace at Deepfields by 1789. He was in partnership with Samuel Pemberton, Benjamin Stokes and Thomas Stokes, initially trading as George Stokes & Co., or Stokes & Pemberton. Coseley Furnace was worked in association with a colliery estate, which previously had been worked by Henry Penn and Michael Pearsall.

The Bickley family were merchants, canal carriers, coalmasters and ironmasters. John Bickley owned mineral estates beside the route of the canal at the time of construction and supplied bricks to contractors in 1769. Coal wharves were established once the canal opened. A carrier's wharf was established at Wolverhampton during the 1780s, first as Bickley & Molineaux, then as Bickley, Whitehouse & Co. and Bickley, Danks & Co. The latter was the most longed-lived partnership that had Benjamin Bickley, Isaiah Danks and Edward Best as members. The names of William and Benjamin Bickley, John Gibbons and Edward Best were included in the Bilston Furnace partnership.

Details of traffic passing through the Dudley Tunnel in the period 1792–1800 reveals the identity of several furnace owners that sent goods and minerals by water. George Parker, at Tipton, and Benjamin Gibbons, at Bilston, both had boats that worked through the tunnel. Wilkinson sent iron from his Bradley Works and Pemberton sent blooms from his Coseley Furnace. Zachariah Parkes also brought goods through the tunnel to his Dudley Port Furnace that included iron ore.

The Parker family had another furnace built near Bloomfield, Tipton, in about 1801 and there was an ironworks and later a second furnace on this site. The Parkers also owned another group of furnaces near the Brades, Oldbury. Profits gained in the trade made the Parkers wealthy. By 1815 control of the assets had passed to the surviving brothers, Benjamin and Abraham, who carried on the coal and ironstone mines, furnaces, forges and ironworks at Tipton, Oldbury and Apedale. Benjamin Parker settled in the High Street, Dudley, with wife Phoebe. He died on 20 July 1816 and parts of his estate were divided between his wife and George Parker, son of Abraham, and George Parker, son of Esther Parker, with George (son of Esther) managing the ironworks. Benjamin Parker's home in Dudley was taken over by Dudley Grammar School.

Rees's Encyclopedia, published serially between 1802 and 1820, included a section on blast furnaces which has been copied and reproduced in other accounts. He quotes six coke blast furnaces in use within Staffordshire, in the year 1788, with another three expected to blow in the same year. The six furnaces would have included three owned by John Wilkinson (Lower and Upper Bradley), two by George Parker (Tipton) and one by Richard and William Croft (Level). The three under construction were probably Bickley's first furnace at Bilston, and Stokes and Pemberton's furnaces at Coseley.

Read, Banks & Dumaresq were responsible for the building of a furnace at Gospel Oak. Stebbing Shaw listed the furnaces and ironworks of Tipton in his book, *The History and Antiquities of Staffordshire*, published in 1801. He shows that the production of iron at that time was quite small:

Messr George Parker and Co. (or brothers) have two furnaces which are estimated to make 25 tons of pig-iron, per week. One rolling and slitting mill, where they sometimes roll boiler plates. Three forge-hammers.

Messr Zachary Parkes and Co. near Dudley Port. One furnace, estimated at 20 or 25 tons per week. One forge. One slitting mill.

Messr Read, Banks and Dumaresq. One furnace, estimated at 25 tons per week, or more. Forge.

Richard Hawkes and Co. One Furnace, lately erected, about 20 tons per week.

Taylor's Foundry at Toll End, where they cast chiefly heavy goods for engines, whimseys and mill-work.

Zachary (or Zachariah) Parkes had a single furnace at Dudley Port, which was built in 1794 on land leased from Lord Dudley and Ward. This furnace was also known as 'Coneygre Furnace' and is not to be confused with another pair of furnaces built across the Birmingham Canal in 1824 for John Turton Ferday.

Read, Banks and Dumaresq's furnace was better known as 'Gospel Oak'. The works was situated beside the old line of the Birmingham Canal and adjoined a large colliery estate containing mines of coal and ironstone. A private branch of the canal, known sometimes as the Dumaresq Branch, passed through the estate to Gospel Oak Canal on the Walsall Level, providing outlets of their products by two separate canal routes. Ironworks and furnaces were erected on the site during 1792 and came to be owned by the partnership of William Bancks, John Read and John Dumaresq. Both Bancks and Dumaresq ceased to be partners from 14 October 1801, when sole charge passed to John Read.

John Read was principally engaged with these works and he also leased the water-powered Gold's Hill Ironworks at West Bromwich (beside the Danks Branch, Walsall Level), Tibbington Colliery, and had shares in other iron and coal works in association with John and Samuel Fereday. His bankruptcy in 1810 eventually led to the protracted sale of Gospel Oak Colliery, furnaces and ironworks and their acquisition by G. and E. Walker.

Richard Hawkes & Co. established their works beside the Walsall Canal at Great Bridge. Here they started to work the local seams of ironstone and coal. There were eventually two blast furnaces, a casting house, iron foundries, stoves, air furnaces, cupolas, refineries, a smith's shop, pattern makers and an engineers fitting-up shop. These furnaces were also known as 'Eagle'. Hawkes also operated a furnace at Graveyard, near Gornal, and during 1811 leased land from Lord Dudley and Ward to erect two furnaces on plot called 'Brecknell', Coseley, in a district also known as 'Deepfields'.

The agreement with Lord Dudley contained the following clause:

> That the said Richard Hawkes his executors, administrators or assigns shall and will on before the twenty fourth day of June one thousand eight hundred and twelve erect and complete and build or cause or procure to be erected completed and built upon said piece or parcel of land or ground hereby demised in good substantial and workmanlike manner two furnaces for making iron with fire engines and other necessary buildings for working said furnaces and shall and will lay out not less than the sum of one thousand pounds in erecting such furnaces and buildings and from time to time carry on and effectually work the said furnaces during the said term hereby demised and without interruption or loss of time save only for such time or times as there shall be an absolute necessity for stopping or discontinuing the working of said furnaces or either of them for the purpose of repairing or for making any necessary alterations therein or by reason of any unavoidable accident that may happen to same.

Richard Hawkes encountered financial difficulties in 1813 and handed over the Coseley Estate to assignees Benjamin Whitehouse & Co., who worked the Coseley Estate until 1816. Attempts to sell these furnaces failed and they remained out of use.

Aris's Gazette published an advertisement in November 1784 mentioning that T. & T. Taylor had taken the Castle Ironworks at Tole End, and they wanted inform factors, ironmongers and others that intended to carry on the business of the iron trade, making engine cylinders, pipes, forge hammers, rolls, presses, stamps, bath stove grates, ovens and all articles in cast iron. Taylor's Foundry is believed to be the same as Daniel Moore's Mill that was located beside the River

Tame at Toll End, and upstream from Golds Hill Mill. These premises were extensively enlarged to become Toll End Ironworks. A long canal basin linked these works with the Walsall Canal. During December 1819 the freehold forge and slitting mill, known as 'Toll End Ironworks', was offered for sale following the partnership ending between Edward Wooley of Bilston and Thomas Pretty of the Toll End Ironworks.

The Tipton group of furnaces also came to include one at Horseley that was made near a private canal built for Dixon, Amphlett & Bedford to serve the Horseley Estate, which they had acquired in 1792. Together they formed the Horseley Colliery Co. Joseph Amphlett, a Dudley ironmonger, provided the practical knowledge, while financial investment came from Edward Dixon, a banker, and William Bedford, a Birmingham solicitor. The canal formed a branch of the Walsall Canal but needed a flight of locks to raise boats to the level of the colliery.

During 1808 the Birmingham Canal Navigations Co. had decided to construct the Toll End Communication Canal, which included the canal belonging to the Horseley Colliery Co. The first Horseley Furnace was built on the Horseley Estate for Harrison, Oliver Kitley and Smith, between 1808 and 1809. They leased a portion of the estate for the purpose of erecting a pair of blast furnaces. Aaron Mamby was also a partner for a time in this venture, which became known as the 'Horseley Iron Company'. Mamby also assisted with the establishment of an engineering works here. The Horseley Iron Co. soon established a reputation for the manufacture of steam engines and structural ironwork.

Iron smelting was carried out at several works near to, or alongside, the Dudley Canal; there was an area at Brierley Hill, known as 'the Level', a surprising name since the land at this point was a definite slope. However, the reason behind the name was not down to local humour, it was because a drain that passed through the land was called the 'Level', and had probably been built to drain mines of the Level Coal and Iron mines.

Within five years of the opening of the Dudley Canal plans were in place to erect a blast furnace to smelt the local ores. Richard and William Croft had leased part of the Level Colliery belonging to Lord Dudley and Ward from 1 January 1784. This agreement allowed the Crofts to build a furnace on the site, which was duly done in about 1786, and the new plant became known as the 'Level Furnace'. A second furnace was added later and a forge was also built. For many years the Gibbons family of Corbyns Hall worked the property.

In 1800 the Gibbons', namely Benjamin, John, Thomas and William, increased their holdings to include mines on the opposite side of Level Street. They elected to build new furnaces on the additional land. These were erected some time after 1806 and eventually four blast furnaces stood on the site. The new group of furnaces became known as 'New Level' and were operated by Benjamin Gibbon junior and Thomas Stokes. The older furnaces and ironworks, now called 'Old Level Furnaces', were re-let to William Izon, while the Gibbons family retained the forge.

The Gibbons were both South Staffordshire bankers and ironmasters, while William Gibbons was a Bristol merchant and iron dealer. During October 1816 the Wolverhampton banking partnership of John, Thomas and Benjamin (the younger) Gibbons faced bankruptcy. This, in turn, affected the fortunes of their ironmaking business. New Level Ironworks and Furnaces were offered for sale. The sales advertisement shows the plant comprised four newly erected blast furnaces, two large foundries, a rolling and slitting mill and a forge. Despite this legal action New Level remained in the hands of the Gibbons family, being worked by Benjamin Gibbons, son of Thomas, and his cousin, William. Fowler's plan (1822) and a book of reference state the occupier of the New Level Foundry, furnaces and ironworks as Benjamin Gibbons junior. Benjamin Gibbons & Co. also retained ownership of the Old Level Forge and Mills.

The mineral estates of Lord Dudley and Ward had been a source of revenue since the times of Edward Sutton, Dud Dudley and their successor Humble Ward, but it was not until the 1780s that their full potential came to be realised. Iron smelting at the Level and Coneygre Furnaces were followed by the erection of another at Parkhead near the south end of Dudley Tunnel (1800–1801), which again was tenanted by Zachariah Parkes and his family.

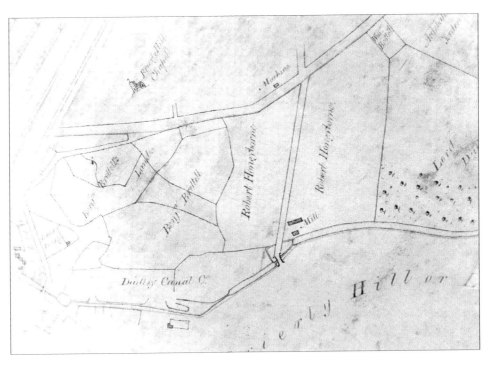

Section of land at Brierley Hill. The plan (*c.*1800) shows Dudley Canal and Brierley Hill flour mill, which was later adapted as an iron and steel works by the Hornblower family.

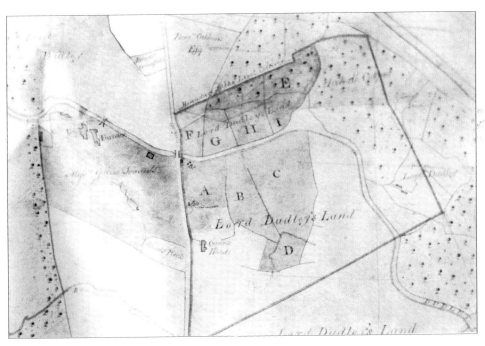

Another section of land to the east of the above (*c.*1800) that shows land leased to the Gibbons family from Lord Dudley, including Old Level Furnaces (left) and the vacant land (C) that was to be developed as New Level Furnaces.

The Parkes family lived in Gornal, where Zachary Parkes senior built a business in auger manufacture. An agreement was made in 1800 between Viscount Dudley and Ward on the one part and Zachariah Parkes, late of Gornal Wood, Zachariah Parkes of Hunts Mill, Josiah Parkes of Gornal Coppice and Zephaniah Parkes of Himley for the building of a blast furnace and lease of minerals. Management of the iron furnaces was the responsibility of Zephaniah, while his brother, Zachariah, continued the auger trade. Zephaniah Parkes & Co. also operated a glassworks at nearby Holly Hall.

Lord Dudley and Ward also arranged for leases that were granted to the Attwood family, who took over land near the Bumble Hole, Darby End and Netherton to erect a group of furnaces known as 'Dudley Wood', the first coming into production about 1802. Four blast furnaces were eventually built there and were sometimes known as 'Netherton Four Furnaces'. They were placed on the south bank of the Dudley Canal and had a tramway link to mines near Corngreaves. The lease was granted in the names of James, Matthias and Aaron Attwood and it was Matthias who it seems had the greatest influence on the concern. Matthias lived at Hawn House, near the River Stour, and had charge of the Corngreaves steelworks beside the Stour. He was engaged in the nail trade and was a partner in the banking firm of Attwood, Spooner & Co. of Birmingham.

Articles of the agreement for the Netherton Furnaces were made during 1799 when the Attwoods were asked to comply with the standard conditions of the time. These included royalty payments for Heathen Coal, Thick Coal, Thick Coal ribs and pillars and ironstone. The method of obtaining ironstone was encouraged to include deep mining. They also had the responsibility of maintaining the 'Level' that drained the mines. Limestone was to be acquired from Lord Ward's mines at a set rate. The supply of coal included mines in Dudley Wood and Bumble Hole, and specific instructions were even given for the building the furnaces. The Grazebrooks were stipulated to be the suppliers of fireclay and firebricks, while the stone for the furnaces was to be obtained from the land or from Messr Bagley at nearby Lodge Farm.

The Grazebrooks were Stourbridge glassmakers who also operated as firebrick makers near Netherton. They also established the Blowers Green blast furnaces, which were located at the end of a private branch of the canal near Parkhead. The partnership in the iron trade involved Michael Grazebrook and Benjamin Whitehouse, whose interests included the Stour Valley Halesowen Forge and Mill, a colliery at Gornal and the Blowers Green blast furnaces. This partnership was dissolved in March 1819 after the death of Michael Grazebrook, with sole control passing to his sons, Michael and Williams, as executors of the Grazebrook Estate. A separate partnership existed for Coseley Colliery, leased in 1810, which existed between Thomas Worrall Grazebrook, Michael Grazebrook and Benjamin Whitehouse. This partnership was also dissolved in March 1819, but the debts were sorted independently between the executors of M. and T.W. Grazebrook and Benjamin Whitehouse and this estate eventually passed to Whitehouse's son, Henry Bickerton Whitehouse. The Glazebrook family retained Blowers Green Furnaces and several generations of this family were counted as ironmasters.

A blast furnace was established by Shropshire ironmasters John Onions and William Bancks alongside the Stourbridge Canal near Brettell Lane, which were known as 'Brierley Hill Furnace'. Bancks and Onions had already established furnaces and mines at Benthall and Broseley. Their works at Benthall commenced during the 1770s and specialised in iron for castings that was sent out by Severn trows to various foundries. Coneybury Furnace, Broseley, was a separate venture established by Bancks and Onions in around 1787.

William Bancks died in 1803 and his sons, William and Christopher, gradually separated their links with John Onions and his son, also John. A new partnership was set up as John Onions & Son and they regularly sent ironstone from Broseley by river and canal to supply the Brierley Hill Furnaces.

The Bancks family, or Banks as they later were known, pursued a number of other ironmaking ventures. They had a furnace in North Staffordshire known as 'Latebrook' and a rolling mill at Brockmoor, they were also associated with the establishment of a furnace at Windmill End and at nearby Netherton. Both William and Christopher Bancks had partnerships in these ventures.

Coalport Bridge over the Severn shows the initials of 'J.O.' – John Onions, ironmaster of Broseley and Brierley Hill.

Windmill End Furnace was located close to the canal near Bumble Hole. A partnership of Christopher Banks, Lathum Blacker and William Benbow had charge of the Windmill End Ironworks until March 1811, when William Benbow left the partnership. Partners changed with regularity and included Theodosia Maddison and William Bancks. By 1817 William and Christopher dissolved the partnership at Windmill End, leaving control of the furnace and a foundry with William Bancks. The furnace was built on land that once belonged to Lord Dudley but had passed to Horace St Paul following his marriage to Anne Marie, daughter of the second Lord Ward. St Paul lived at Ewart Castle in Northumberland and had titles that included Baron and Knight of the Holy Roman Empire. St Paul and his descendents also gained other mineral properties that included Tipton Forge and Willingsworth Furnaces. Christopher Bancks was also connected with Netherton Furnace, erected near Bumble Hole. Archive records suggest that William Bancks leased the Netherton Estate from Lord Dudley in 1813. Lewis's *Trade Directory* quotes Christopher Bancks as owner, but within a year Netherton Furnaces were taken over by the Attwood family and were worked together with their other four furnaces at Dudley Wood.

An extensive mineral estate existed at the Buffery, which was served by tramways to the Dudley Canal. Richard Salisbury and Fereday & Co. both established blast furnaces that became respectively known as 'Old' and 'New Buffery'. Richard Salisbury was a Belfast merchant who moved to Dudley to manage the estate. A lease was agreed between the Hawkes family, owners of the core mineral estate, Richard Salisbury of Belfast, Paul Tate of London, Francis Hone, a captain in the 3rd Dragoon Guards and Robert Keate of Mayfair, London. A partnership was formed in August 1813 to work the estate that comprised Richard Salisbury (merchant), Thomas Hawkes (glass merchant), Robert Keat (surgeon) and Paul Tate. Several lots of adjacent property were acquired and the Old Buffery Furnaces and Ironworks were established.

Samuel Fereday, John Jones and James Wainwright owned the New Buffery Furnaces and Colliery, which were located to the east of Salisbury's estate. A branch canal to the Bufferies was contemplated when the Dudley No.2 Canal, or Netherton Canal, as it was sometimes known, was constructed. Many locks would have been needed to build this canal and the scheme did not proceed. Tramways and roads provided the only means of conveying iron to the canal. Salisbury's tramway travelled west towards the canal at Blowers Green, while Fereday's tramway travelled south to wharves at the Bumble Hole.

There were a group of furnaces located around Gornal that had no canal access but benefited from local reserves of both coal and iron. John and Edward Crockitt worked Dibdale Furnace until their bankruptcy in 1819. During February 1820 the Didbdale Furnace with engine, warehouse, carpenter's shop and blacksmith's shop was advertised for sale. Gornal Wood Furnace and mines belonged to William Hornblower, who was also associated with the Brierley Hill Iron and Steelworks at the Delph. The third furnace was known as 'Graveyard', which was worked by Thomas Jones until his bankruptcy.

The Addenbrookes were a family of Stourbridge ironmasters, who operated Broadwaters Forge near Kidderminster and owned the Royal Forge on the Stour north of Stourbridge. They were also responsible for the setting up of the Moorcroft Furnaces near Wednesbury. Constructed in around 1801, on land leased from 1800, these works were located beside the private Scott & Foley owned Bradley Branch Canal. Tramways fanned out from these furnaces to collect coal and ironstone from the adjacent colliery estate that belonged to Scott & Foley.

Proprietors for the Moorcroft Furnaces were brothers Edward, Henry and John Addenbrooke, and each sibling also had the middle name Addenbrooke. The reason for this repetition was due a change of name. Their father was ironmaster John Addenbrooke Homfray who was partner in the Lightmoor Furnaces in Shropshire with Francis Homfray. John Homfray was the chief beneficiary of the will of Edward Addenbrooke of Wollaston Hall, Worcester. On the death of Edward Addenbrooke of Wollaston the estate was bequeathed to relative John Addenbrooke Homfray on the condition that he assumed the name and arms of Addenbrooke. The Homfrays were important ironmasters in their own right and Francis Homfray was associated with the foundation of the Penydarren Ironworks, near Merthyr Tydfil, South Wales, in 1784, and which the Homfray family retained until 1822. Moorcroft was one of a small nucleus of coke-smelting furnaces that grew up around the Walsall Canal in the Wednesbury area. The others were the Wednesbury Ironworks, owned by the Hallen family, and Leabrook Furnace, owned by Michael Toney.

Wednesbury Ironworks comprised a blast furnace, foundry, boring mills, slitting and rolling mills, pattern and smith's shops that were located alongside the Monway Branch of the Birmingham Canal. The owners were John and Samuel Hallen who were made bankrupt in 1796. Associated with these works were mines of coal and ironstone at Great Bridge and Moxley.

John Bradley formed an important iron-working partnership at Stourbridge with the Foster family, when he moved there from Shropshire in around 1798. Initially Bradley & Co. were makers of iron goods and tools such as spades. Financial assistance was provided by Henry Foster of Nantwich, a woollen draper. When Henry died, John Bradley formed a new partnership in 1802 that comprised Bradley, Thomas Jukes Collier of Wellington (wine merchant), Thomas and James Foster of Alvaston, William Foster of Nantwich (woollen draper) and the children of Henry Foster, deceased, that is Elizabeth Foster of Nantwich and James, Lucy, Mary and William Foster of Stourbridge. Their ironworks was located on a strip of land placed between the Stourbridge Canal and the River Stour and adjoining the Royal Forge owned by John A. Addenbrooke. The younger James Foster (of Stourbridge) was employed as a clerk in the Stourbridge works, but later came to be a partner in the firm.

The Wyrley & Essington Canal passed to the north of Walsall and although there were iron mines, only one group of ironmasters chose to mine there. Stubbs and James established the Birchills coal and ironstone mines and erected a blast furnace there. In 1815 when these works went on sale, the estate included a mine engine, whimseys, a blast furnace cased with iron plates, a blast engine casting house, a new foundry, a clay mill, sheds and brick kilns.

Thomas Price built an integrated business that utilised the Walsall Canal. Limestone mines were leased from the Earl of Bradford in Wolverhampton Street, Walsall. Ironstone mines were worked at James Bridge, while blast furnaces were established at Bilston Brook at the terminus of the Bilston Branch. Price occupied a mineral estate adjacent to the furnaces that extended from the Bilston Branch towards the line of the Old Birmingham Canal near Glasshouse Bridge, where Thomas Price had a canal basin.

Development of the Puddling Process

The Merthyr Tydfil area in South Wales became the cradle of a significant development in ironmaking known as the 'puddling process', or 'Welsh method'. Prior to the development of this method ironmasters were restricted to hammering their pig iron to make it malleable. Ironmasters such as the Cranage family at Coalbrookdale and Peter Onions (at Broseley and later Dowlais) had experimented with other techniques, but had not found a successful process. It was left to Henry Cort to make the essential breakthrough after trials at his water-powered forge and slitting mills at Fontley, near Fareham, between 1783 and 1784. Essentially he introduced a two-stage process where the reverberatory furnace was used as a means of working the iron and grooved rolls were introduced for rolling out the metal as sheet, strip, etc. While initial success was made at Fontley, the process needed refinement and Richard Crawshay achieved this at his forge at Cyfarthfa, Merthyr Tydfil.

Cort's invention was, however, a crucial step. Few inventors have made such a significant contribution to the iron trade as Cort did for so little reward. He was made bankrupt in 1789, partly through money spent on perfecting the method and partly through the insolvency of his partner, Adam Jellicoe, who had died.

Those who tried to duplicate Cort's method encountered problems with certain types of pig iron, especially the 'grey forge' version. The method devised was to heat the iron in a finery, run out the molten iron into a trough and separate the iron from impurities that formed as slag. This technique became known as the 'running out fire' and the purified iron went to the puddlers. Some sources credit Joseph Firmstone with the development of this method while working at Dowlais, others credit Samuel Homfray at Penydarren. It seems to have been adapted across the country. Advertisements for the sale of the Wrens Nest Forges, Shropshire, in 1804, mention the use of a running out fire connected with the finery.

With the introduction of puddling came enlarged infrastructure. Puddling furnaces produced metal known as 'puddled ball'. This intermediate stage was achieved after the pig iron was broken up and melted in the reverberatory furnace. The puddler's task was to stir and mix the iron while controlling the heat. The process was a skilled, and also physically demanding, task. Puddlers needed strength to turn and stir the metal until it reached the correct consistency. This was done in a continuous heat. Eventually the iron reached the state when it was rolled up into a ball and removed from the furnace. The ball was carried to the shinglers hammer to be stamped to expel the slag and shaped into a billet. From there it passed to the rolls. Successive reheating in furnaces and passes through the rolls first produced what was termed 'merchant bar'. Piling the bars, as in the previous method of the finery and chafery, and further rolling led to higher quality brands. With the additional stages, opportunity for improvement was ever present and ironmasters frequently strove to make alterations and changes to the basic process.

Puddling furnaces and rolling mills became standard features in the iron trade. In South Staffordshire a pioneer of this method was Joseph Firmstone who brought the technique from South Wales. Through the medium of the puddling furnace, iron was converted into malleable, or wrought, iron. In this form the iron became more pliable, it could be rolled and shaped more easily and the number of uses increased tremendously. Early references to puddling include the Nine Locks Forge, Brierley Hill, that *Aris's Gazette*, in April 1802, reported to have 'blast and balling furnaces', and the Leabrook Ironworks, Wednesbury, that in December 1803 had puddling furnaces and an 'air chafery'.

Brockamoor Forge was another ironworks with puddling furnaces. The plant as described in December 1803 comprised two puddling furnaces, a balling furnace, an air chafery, a running out finery and hollow fire and was capable of making between 40-50 tons of iron per week. John Wilkinson was a keen supporter of the technique and had installed puddling furnaces at Bradley by 1805. They subsequently became a standard feature in local ironworks in association with the rolling and slitting mill.

Another crucial innovation to the iron trade was the application of the steam engine to drive the rolling mill, which gave ironmasters the ability to move iron working away from the watermill. Hitherto this development, ironmasters sought out water-powered mills to roll bar iron, slit or draw into wire. The transition to using steam engines took place over a period of years, as watermills continued to have a role in the trade.

Certain watermills such as Broadwaters, Gothersley, Heath, Hyde, Kinver and Whittington continued to perform an important role in the local trade. They adapted with the industry, utilising both steam engines to turn the mills and puddling furnaces for the production of wrought iron.

Samuel Fereday, Ironmaster

Samuel Fereday was the most important Black Country industrialist in the years between the death of John Wilkinson in 1808 and the post-war slump of 1816. At the height of his prosperity he had shares in blast furnaces, ironstone mines and collieries and employed some 5,000 workers in mines and ironworks.

Samuel Fereday was the son of Samuel Fereday, stonecutter, stonemason and quarry master of Upper Gornall. The younger Samuel Fereday married Ann Bagley, daughter of ironmaster Dudley Bagley, whose works included Lye Mill. Samuel and Ann had three children, Ann, Sarah and Dudley. Samuel Fereday began as a farmer at Park Farm, but extended his commitments to include the coal and iron trade; it was fortunate that his farm estate contained ironstone and coalmines. His farmhouse home became known as 'Ettingshall Park' and it was from here that he ran his affairs. John, second son of Samuel Fereday senior, married Jane Turton and continued his father's business as quarry master. They had a son, John Turton Fereday, who, like his uncle, found profit in the iron trade.

The Ettingshall Park and Parkfield Colliery and Ironstone Works were a vast, untapped mineral estate developed by Fereday & Co., where the partnership was Samuel Fereday, John Read, Richard Smith and George and Thomas Stokes. In July 1813 three of the partners, John Read, Thomas and George Stokes, left due to financial reasons. George Stokes was replaced by assignees Thomas Hampton, Joseph Hancox and George Talbot, while Thomas Stokes was replaced by his assignees, John Jeffreys, William Stevens and Thomas Tickell.

Fereday & Co. partnership, Furnaces 1812

Bradley (Lower)	Fereday, Bickley & Smith
Bradley (Upper)	Fereday, Bickley & Smith
Buffery (New)	Fereday, Jones & Wainwright
Hallfields	Fereday, Bickley & Smith
Millfields	Fereday, Turton and Walker
Priestfield	Fereday, Smith, Ward
Roughills	Fereday & Co.
Tipton	Turton & Fereday

John Turton Fereday, William Turton and John Parsons Firmstone were partners in the Highfields Mills and Tipton Forge from 1809, which was the same year that a new group of furnaces were constructed at Tipton Green. The partners in the Tipton venture were William Turton, Samuel Fereday and his nephew, John Turton Fereday. The furnaces were built into the side of the bank that led up to the Old Main Line and faced a private arm made to join up with the Toll End Communication Canal. Tipton Forge supplied blooms of iron to be worked at Highfields Ironworks. This arrangement continued until 1820 when the partnership was dissolved.

The name of Firmstone was long associated with the West Midlands iron trade. The first member of the family to settle in the region was Joseph Firmstone. Joseph married Margaret Parsons at Broseley in April 1783. Their union produced five sons: George, John Parsons, Joseph, Thomas and William who all became associated with the coal and iron trades. Joseph Firmstone senior moved to South Staffordshire where he became a partner in the Highfields Colliery and Highfields Ironworks. Highfields Colliery was an estate divided into thirty-two shares, with shareholders including Joseph Tarratt, Benjamin Stanley, Joseph Firmstone, William Sparrow, Thomas Timmins and William Parsons. The ironworks had been started by 1804, when the partnership comprised Thomas Ellis, Joseph Firmstone, William Parsons and Thomas Perry. During December 1804 the partners decided to trade under the name of William Parsons & Co. The ironworks comprised a steam-driven slitting mill for nail rods, a rolling mill for bar iron and a foundry.

The working arrangements at Highfields were somewhat complex with different partnerships associated with the colliery and ironworks. In around 1806 John Parsons Firmstone joined the Highfields Ironworks partnership and this arrangement lasted until about 1 November 1809.

Some time before the termination of th Highfields Ironworks partnership, negotiations were made with Samuel Fereday and William Turton for the lease of the Highfields Ironworks, but they refused to rent them until J.P. Firmstone agreed to join them and become a partner. It was afterwards arranged with William Parsons, Joseph Firmstone senior and Thomas Perry that John Parsons Firmstone should become a partner with Samuel Fereday and William Turton in manufacturing iron at Highfields. J.P. Firmstone refused so to do without said Parsons, Firmstone and Perry, guaranteeing him from loss. For the Fereday organisation, the Highfields Ironworks was yet another addition to a rapidly increasing industrial empire, which at this time also included the recently acquired Bradley Ironworks, formerly John Wilkinson.

Samuel Fereday, at his home, Ettingshall Park, had two eligible daughters, Elizabeth and Sarah. Richard Smith married Elizabeth Fereday at Sedgley parish church on 12 June 1811. Their first child, George Samuel Fereday Smith, was born on 7 May 1812 and a daughter, Sarah, followed in 1814. After ten years, Richard and Elizabeth had a second son, Frederick, in 1824.

Richard Smith was a very capable person and no doubt his father-in-law saw promise in the young man. Richard Smith was born on 30 January 1783 into a family long-connected with coalmining. His father, Thomas Smith (1747–1808), is said to have been a working collier, but may have had a more important role such as a ground bailiff, or a 'butty'. Thomas came to Deepfield, near Coseley, from Brierley Hill in around 1776 and rose his status to coalmaster and then to coal and ironmaster. With the attendant prosperity Thomas Smith was able to make Tibbington House, near Princess End, his home.

Tibbington House had been bought from Francis Stokes. An estate plan of 1817 shows the house, garden and surrounding fields. This plan also illustrated the Tibbington Ironworks, a whimsey and the pits of Tibbington Colliery and a canal basin opening onto the Birmingham Canal. The minerals were leased to John Read and were offered for sale in December 1817. Four pits, drawing both and coal and ironstone, were then in work that were served by a private arm of the Birmingham Canal. Tibbington House, then in the occupation of Richard Smith, covered an area of 2 acres, three roods and twenty perches, and in addition to the house there was a garden, plantations and pleasure grounds!

Thomas Smith and his wife, Mary, had nine children that survived infancy: five sons and four daughters. Richard Smith was the fourth son. Richard grew up in a strongly religious household; his father was a strict Baptist. In the year of Richard's birth, Thomas Smith and others had separated from the Dudley Baptist church and had begun meetings at the Darkhouse, Coseley. The origin of the name 'Dark House' is attributed to the wood of surrounding trees which obscured the light. In 1787 Thomas Smith became a pastor at the new chapel, erected there in 1785, and continued in this role until his death twenty-one years later. Richard later erected tablet at Darkhouse chapel to his father's memory.

Thomas died a few days before Christmas 1808, when he succumbed finally to the painful illness that had afflicted him. At the time of his death he owned 240 acres of land containing the thick, or Ten-Yard Coal, at such places as Capponfield, Deepfield, The Dimmocks at Brierley, near Bilston, and Tibbington House. He also leased land for mining at Moat Farm and elsewhere. Much of the property was owned or worked in partnership. Thus, Thomas Smith worked with the Stokes at Deepfield, bought the Dimmocks jointly with Samuel Fereday and Thomas Price, and leased Moat House Farm in partnership with William Underhill, members of the Fereday family and others. There were blast furnaces erected on the Capponfield Estate that Smith and John Read had shares in.

Richard Smith left school and began work with his father and acquired a considerable ability in mining matters. The local iron and coal industries continued to draw on war department orders and this encouraged a rapid increase in the number of furnaces. The demand for iron continued to increase and the price was maintained through to 1811. It was a time when over 8 tons of coal, 2½ tons of iron ore and nearly a ton of limestone were required to make a single ton of pig iron, and further large amounts of coal were subsequently needed to transform the pig into bar iron – coalmasters and ironmasters prospered together.

Profit was inevitably accompanied by risks. These were times where iron was in demand for ordnance. The war with France had thrown a greater dependence on the British ironmaster and reduced the need for imported Swedish and Russian iron that at one time had been an important supply for the ironmasters working up pig or bar iron into finished products. The alarming prospect of invasion by Napoleon's army led to the formation of local volunteer regiments and a new requirement for arms and armaments. The concern first surfaced during 1803 and was met with countrywide support. Committees were formed and the volunteers were formed into companies, battalions and regiments. Staffordshire eventually had four regiments. Both Richard Smith and his cousin, John Turton Fereday, gained commissions as captains of companies in the West Staffordshire regiment.

When his father died in 1808, Richard was named as one of the executors and his share of the estate was worth several thousand pounds. Henceforward, Richard and his brothers were independent coal and ironmasters involved with such men as George and Thomas Stokes, the Gibbons family and the Feredays in the usual varied ironmaster partnerships. The links with Samuel Fereday and his associates were soon to be the strongest. Both Samuel Fereday and his younger brother, John, had been partners and later executors of Richard Smith's father, therefore they knew Richard well.

This was a period when Fereday was adding to his holdings; with the death of John Wilkinson in 1808 he was able include the Bradley Collieries, Ironworks and Furnaces in his industrial empire. Richard Smith clearly prospered from the relationship. He became partner with his father-in-law and Benjamin Bickley in the Bradley group of Ironworks. He also joined Fereday in the partnership of Priestfield Furnaces and numerous collieries. On a financial level, Richard Smith, Samuel Fereday and James Fisher became partners in the Bilston Bank, and also with Fereday and Samuel Wagstaff became partners in the Kidderminster and Stourport Bank.

Family relationships and friendships were the backbone of the local iron trade during this period. Unfortunately, the trade was soon to face some severe tests and the many links forged in these partnerships would reach breaking point.

CHAPTER 3

MISFORTUNE AND UNREST
AN INDUSTRY IN STAGNATION, 1814–1826

Pig iron prices dropped sharply in 1812 to £5 10s a ton and the fall continued the following year. By 1816, at the post-war slump, pig iron was priced at £3 15s a ton. Samuel Fereday's first response to the sag in prices was to issue two million trade tokens in the wages of his 5,000 employees during 1811 and 1812. There was some justification for the issue; copper coins were in short supply – but the tokens were so lightweight and the resulting outcry led eventually to their suppression. In November 1815 there was a run on the bank of Fereday, Smith and Fisher. As unemployment grew, disorder occurred; special constables were sworn in and the Yeomany Cavalry was used to disperse the crowds.

Considerable financial pressure was now put on Fereday and his remaining business colleagues. He had lost the support of John Read and Thomas Stokes through bankruptcy and Dudley Bagley, his father-in-law, had also faced the bankruptcy courts. Some of Fereday's partnerships began to be dissolved by mutual consent, but more troubled times lay ahead. On 30 July 1816 Richard Smith was made bankrupt.

The Gibbons family also started to give up their property in the Bilston district. They had been associated with the Bilston Furnaces (Bickley, Gibbons and Best), Bilston Ironworks and the Cockshutts ironstone mines. During November 1816 Thomas Gibbons, John Gibbons and Benjamin Gibbons junior arranged for the sale of the coal and ironworks at Cockshutts, belonging to the estate of the Earl of Darlington, and also for their contract with George Rushbury for working the estate. In January 1817 they advertised the sale of Bilston Mill, which was quite extensive by the standards of the time. It included a forge and two iron helves (worked by an engine with a 26in cylinder), fourteen puddling furnaces, a mill to roll hoops and bars, slit rods, a small round or flat iron (driven by a Boulton and Watt steam engine with 42in cylinder) and six heating furnaces. The whole plant was capable of making 120 tons of iron per week.

Brothers William Hanbury and John Sly Sparrow took over Bilston Mill and the operation of the adjacent blast furnaces, which was the start of a family connection with Bilston that lasted through to the 1880s. The Sparrow family, in South Staffordshire, had originated in Wolstanton, North Staffordshire, with family members that included Burslem, John and William. They had several links with the iron and coal trades in the district, such as Sparrow's Forge on the Tame, near Wednesbury, a partnership with Bilston coalmaster Thomas Tomkys at mines in Bradley as well as other mines in their own name in the Ettingshall and Sedgley area. William Sparrow married Mary Mander, the daughter of Thomas Mander of Birmingham. Their sons William and John were to make an important contribution to the iron trade. William Hanbury Sparrow (1789–1867) and John Sly Sparrow (1790–1825) were associated with Samuel Fereday in a colliery at Bilston from about 1812. They then went into the iron trade with the acquisition of the Bilston Furnace during 1816, this was followed by the purchase of Bilston Mill. John Walker, ironmaster, was a partner in their business until 1820. For the Sparrow brothers there were difficult times ahead, but even a declining iron trade did not deter them from investing in new ventures.

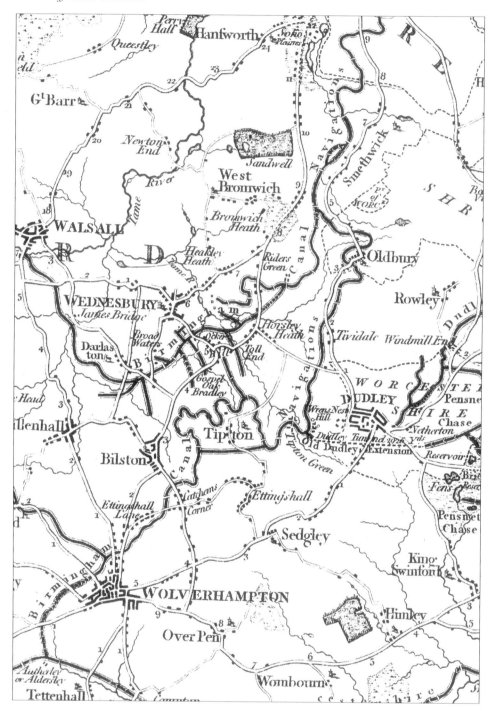

Carey's map (1796) shows the canals and turnpikes around Bilston, Tipton and Wednesbury. These roads and waterways served the heart of the fledgling iron industry. Carey's map is accurate with the exception of two items. The flight of locks to the right of Dudley (at the Buffery) was not built. This line of waterway was made as a tramway to avoid excessive lockage on the recommendation of Thomas Dadford (senior), engineer. The other item was the link south of Tipton, which was the line of Coseley Tunnel. This link was not complete until 1838.

Slump in the Iron Trade

After 1815, with the end of the Napoleonic Wars, the iron trade went into a severe decline, as less iron was needed during peace than in wartime. Many important ironworks and collieries were closed. The Fereday family suffered badly; Samuel Fereday's industrial empire collapsed after the Battle of Waterloo. The fall in requirements for iron ordnance had a serious effect on the domestic iron industry. The slump had been coming for around four years.

Manufacturers sought fresh means to counteract the slump. Perhaps the most reprehensible was the insidious truck system method. It was a practice where manufacturers gave payment for work in the form of goods from the company shop. The shops were commonly known as 'Tommy Shops' and some owners operated these ventures for the benefit of their workers. Unfortunately, there was an element that saw these shops as a means of control and abuse. Substandard goods and food were sold and the men and their families were obliged to purchase them, failure to do so would result in loss of work. Associated with the shops were houses that sold beer; workers were also encouraged to spend their hard-earned wages on alcohol. In both cases, the workers and their families suffered. The practice became common throughout South Staffordshire and affected many sections of the labouring community and especially the chain makers, ironworkers, miners and nailers.

The slump caused a considerable number of miners and other workmen to lose their jobs. Ironmasters began to lay off men from November 1815 and reduced the wages of others. Groups of miners congregated in places such as Bilston, Coseley, Dudley and Wolverhampton and some began roaming around the district armed with clubs and sticks. Soldiers and the calvary were drafted in from different places including the Staffordshire Yeomany, Leek Yeomany Cavalry, Manchester Ninth Light Dragoons and the Middlesex Militia to keep order and disperse groups of miners. Their presence curtailed much of the violence although there were some serious incidents. Zephaniah Parkes was attacked near Holly Hall, Dudley, in January 1816. Rioters put a rope around his neck and threatened to hang him unless he agreed to their demands. Some demonstrations were more peaceful; during 1816 a group of Staffordshire miners decided to make their plight known to the government by yoking themselves to several wagons laden with coal and armed with a petition addressed to the Prince Regent. They carried placards which read, 'Rather Work than Beg'. They travelled in different groups to the Metropolis. Two groups were stopped by 'police' officers at St Albans and Maidenhead and were persuaded to return home, their coals being distributed amongst the poor of those towns.

Unfortunately, such was the state of the trade that unemployment and dissent amongst the workforce would last for a number of years. Unemployed miners would roam the countryside begging for food. Some found employment working on the harvest during August and September and there were suggestions that out of work miners might be employed in improving roads around the parishes. Regrettably, the means of funding road improvement was through raising highway rates and because of the slump in trade generally this idea was faced with opposition. Eventually those out of work drifted away to jobs elsewhere. The unlucky, that fell foul of the stringent legal system that then prevailed, could face transportation, or worse. The death penalty included burglary and forgery at this time and the Assizes at Stafford, Warwick and Worcester were busy places for those who transgressed.

Towards the end of 1815 Samuel Fereday started to reduce his share holding in the core ironworks under his control. This did not stave off bankruptcy, however, when in April 1816 Fereday, Fisher and Smith, bankers of Bilston, went before the bankruptcy court. Richard Smith also faced bankruptcy in his own right. The toll of bankruptcies that had begun with John Read and followed with George and Thomas Stokes, had now also claimed Samuel Fereday and Richard Smith. Their affairs were now in the hands of the creditors, lawyers and judges. Proceedings could run on for years and many lives were affected. As creditors closed in, another bankruptcy order was placed (in 1821) against Samuel Fereday, solely in his own name. Meanwhile, Fereday left England and went to manage an ironworks near Paris, thus escaping the harshest moments that bankruptcy brought. Samuel failed to make a success of this venture and retired to Capelle, near Boulogne, where he died in 1839. Richard Smith stayed on to deal with his own cause and the cause of his father-in-law, working in the trade as best he could to pay off his debts.

While the young Richard Smith fought an uphill battle with creditors, considerable respect was earned through his practical knowledge of mining. He acted as a mineral surveyor, producing plans that assisted with the development and sale of various mining properties. He worked from his home at Tibbington House until 1825, and it was here that his son, Frederick Smith, was born. By 1826 Smith had moved to London, where he earned his living as a mineral surveyor. He went to Portugal that same year to consult on a mine and returned to London in 1827 to receive another job from Rundell & Co., jewellers, gold and silversmiths to the King, who had obtained mining rights for Nova Scotia. Their lease was exploited by a subsidiary of Rundells, the General Mining Association, and Richard Smith was given the task of managing the operation. Richard Smith left with his family for America and it would some nine years before he returned.

The break-up of the Fereday concern was piecemeal. Some parts had been deployed to friends and other family members. It was in 1815 that the Bickley, Fereday and Smith partnership at Bradley was taken on by nephew, John Turton Fereday (1786–1849). In July 1819 the respective one-third shares owned by Samuel Fereday and Richard Smith in the Priestfield Ironworks came up for sale. Priestfield at this time included three blast furnaces, a water engine, foundries, casting house refineries and mines of coal and ironstone. This venture was established in 1808 and included William Ward as a partner. Ward was to gain complete control of the Priestfield concern.

Samuel Fereday, William Turton and J.P. Firmstone traded as the Highfields Iron Co. until 1817 when Samuel Fereday left the partnership. William Turton and John Firmstone were left in charge. It unfortunately happened that in the closing of the accounts of Fereday, Turton and Firmstone, J.P. Firmstone had incurred losses amounting to over £4,000 through his share, but managed to recoup some £2,500.

A distinction between the working company, the Highfields Iron Co., and the property owners must be made. Ownership of the ironworks and collieries was principally with the Firmstone family. By 1810 the ownership of the Highfields Ironworks was divided into three shares with William Parsons, Joseph Firmstone, and Thomas Perry and J.P. Firmstone owning a third. Working the Highfield Ironworks came to be divided between two concerns. One agreement led to the letting of the rod side to William Turton and the bar side to John Walker.

Working agreements were fluid and changed from year to year. Specific arrangements were made as to specific ironmaking tasks. An agreement made in January 1819 between J.P. Firmstone and William Firmstone, ironmasters, referred to J.P. Firmstone's interest in Highfields Mill and Ironworks. J.P. Firmstone being then associated with William Turton at Tipton Forge and Highfields and having made arrangements with Turton to manufacture for him (J.P. Firmstone) 30 tons per week, on average, bar and rod iron or 20 tons per week of hoop iron. J.P. Firmstone agreed to manufacture for J.P. Firmstone and William Firmstone a similar quantity of finished iron. J. and W. Firmstone, trading as makers of pig iron (at Roughills Furnaces), agreed to deliver to Tipton Forge as much pig iron as will amount to the bar, rod or hoop iron to be manufactured.

The location of Tipton Forge has been traced to the Bloomfield district of Tipton, near the Birmingham Canal, and was later known as 'Factory Forge', or 'Factory Ironworks'. Movement of iron would have principally been by canal, with boats making the trip along the canal between Highfields Furnaces, Highfields Ironworks and Tipton Forge. In these early days of steam-powered works, the concept of integration had not been adopted and little concern was given to the cost of carriage.

Another agreement was in force between the Highfield Iron Co. and George Firmstone and John Cornforth to manufacture bar rod and other kinds of iron. The Highfields Iron Co. agreed to deliver to G. Firmstone & Co. common and best blooms longweight of sufficient quality to make good bar iron and rod iron at £11 per ton on the wharf and plate metal of good quality at £7 15s per ton.

Costings and charges were complicated by the method of weighing known as 'longweight' and 'shortweight'. In the former, allowances were made for loss and a ton might vary from the equivalent upward from 21-24cwt depending on the allowances made. This practice made it very difficult for the canal companies who calculated their tolls on the basis of distance and weight. Accurate gauging of boats would result in the boat owner paying the equivalent of higher tolls per ton; resulting complaints from carriers ensured latitude being given by the local canal companies, at least.

For Firmstone and Cornforth a detailed set of charges was agreed. Castings such as rolls and pinions made from good pigs cost £12 per ton and open sand castings cost £8 per ton on the wharf. Coals into the basin at 7s 6d per ton longweight, lumps into the basin at 6s, slack into the basin at 2s 6d longweight, coals into the works where they may be wanted at 8s per ton, lumps at 6s 6d per ton and slack at 3s per ton. Highfields Iron Co. was, however, not to deliver any coal or slack further than the basin. G. Firmstone & Co. also agreed to deliver various grades of iron to Highfields Iron Co.'s warehouse or yard:

Common merchant bars with ends sheared off at 12 pound nine shillings per ton *shortweight*, common rod iron, bundled with either three or four bands at 13 pounds nine shilling per ton *shortweight* to wharf

Common rolled iron from 2 inches upwards to four and a half inches broad and not thinner than no: 14 wire gauge at 13 pounds per ton

Common merchant iron 3/8 and 7/16 round at 16 pounds ten shillings per ton *shortweight*

Common bars ½in and 9/16in round at 16 pounds ten shillings per ton *shortweight*

Common ½ square at 13-10s *short weight*

Best rod iron all sizes at 13-9 per ton except when he rods are picked or any cut from them or when bundled with hot bands when G.Firmstone and Co. are to pay five shillings per ton extra allowed

Merchant bar iron all sizes as low as 1¼in wide made from the plate metal at 12-19-6d per ton *short weight*

Rod iron made from the plate metal at 14-1s per ton

Common merchant bar iron from blooms from one inch broad to 1¼in at 14 pounds per ton short weight if made from the plate metal to be allowed the difference in price between that and blooms

The above list demonstrates the already complex arrangements that existed for iron manufacture and especially how different aspects of the process were contracted for. The end products were different grades of iron (common or best) made as hoops, bars or rods that passed to a central warehouse for distribution. The traffic along roads and canals, back and forth, of minerals, pig iron and finished iron was then accepted with apparent disregard of the transport costs involved, the guiding factor being where business could be conducted rather than consideration for a central manufacturing location.

Other people formerly linked with the Fereday business also found this a difficult time. George Stokes, nail ironmonger and ironmaster of Oldswinford and Kinver Mill, had been one of the partners in Parkfield Colliery, Wolverhampton, the Deepfields Furnaces at Coseley, New Coseley Colliery and the owner of mines at Rowley Regis and Netherton and was also concerned with a mineral estate owned by Lord Darlington at Billingsley in Shropshire. Both coal and ironstone was mined on this estate and two blast furnaces were established to smelt the local ironstone. A railway, some two miles long, was laid with cast-iron rails and sleepers to connect mines and furnaces with the River Severn. An incline plane brought wagons down to the riverside wharf. George Stokes was made bankrupt in 1813 and the various estates passed to his executors who gradually disposed of his holdings. Billingsley Mines and Furnaces were advertised for sale in 1818.

John Turton Fereday carried on the family name by continuing some of the partnerships established by his uncle. He was associated with the Tipton Furnaces, mined iron ore from the Roughills Estate and worked the Upper Bradley Furnaces and Ironworks. During 1817 he opened up new coalmines at Dudley Port and in 1824 he also arranged for the building of two new blast furnaces there, which became known as 'Dudley Port Furnaces'.

Ironmasters Organisation

In a trade formerly dominated by a select group of families, these troubled times encouraged the ironmasters to work together. They started to hold quarterly meetings in different towns of South Staffordshire to deal with problems common to all and eventually set rates for the different grades of finished iron. A prelude to this working association was a meeting of ironstone sellers held at the *Swan Hotel*, Wolverhampton, on 9 March 1818, where it was resolved to set the prices for ironstone delivered into boats for the next quarter.

An account of their resolution was recorded in a document, preserved at Wolverhampton Archives (DX 84/21).

Ironstone to be gotten	Tons	*Ironstone wanted*	Tons
R. Smith	40	J. & W. Firmstone	320
J. & W. Firmstone	380	J. Walker	340
Grazebrook & Co.	120	Priestfields	400
Timmins & Co.	220	Bilstone	420
Sparrow & Sons	400	Caponfield	350
J.F. Fereday	600	Deepfield	500
T. Smith	450	Bradley	550
W. Ward	450	Addenbrook	300
J. Walker	400	W. Oak	300
Philpot	40	G. Oak	300
G. Jones	800	J. Bagnall	170
G. Beaumont	40	Horseley	300
Bagnall & Co.	300	Birmingham Coal Co.	370
Williams, Jones & Co.	100	Tipton	370
Penns Fields	120	Parker	60
Stevens & Wallbrook	300	Levell & Banks	300
Granger	30		
Waterhouse	40		
Hill	60		
Foxyards	200		
Tipton Furnace Co.	120		
Elliott	30		
Addenbrook	50		
Birmingham Co.	150		
Horseley Co.	200		
J. Fellows	50		
Hawkes	80		
Bow Dawn	80		
Hunt	40		
Total	**5,920**		**5,350**

Prices were set at the following per ton: New Minestone, 12s 6d; Whitestone, 13s; Top Gubbins, 13s; Blue Flats, 14s, Deep Gubbins, 14s, Ballstone, 14s and Poor Robins, 14s – although reductions were made for immediate payment. Those who signed the agreement were J. and W. Firmstone,

This view of Bradley is reproduced from the Centenary Booklet produced by the Staffordshire Iron & Steel Institute in 1966 and mentions the date of 1836. The artist was Robert Noyes whose painting has been stated to be from 1817. Contemporary map evidence would seem to indicate that the Wilkinson ironworks of Hallfield, Upper Bradley and Lower Bradley were arranfged along the skyline in this view. The nearer furnace appears to be Moorcroft, owned by the Addenbrooke brothers.

W. Grazebrook, William Hunt & Sons, George Jones, W. Sparrow & Sons, Thomas Smith, Thomas Timmins & Co., William Ward, John Walker and Phillip Williams & Co.

This document also included a list of ironstone production and requirements that is worth reproduction. Many of the principal furnaces and their requirements are given. The largest ironstone producer is listed as George Jones, who at this time did not smelt ironstone, while Bagnall & Co. and J. T. Fereday belonged to a group that produced stone in excess of their requirements.

Bankruptcy and Unrest

The Government of this period showed a degree of insensitiveness when they tried to impose a tax on coal during 1819. Meetings were held throughout the region and deputations were appointed to lobby Members of Parliament. Stagnation in the iron and coal trades within the previous two years was most marked, but the chief concern from local coal and ironmasters was that the price of manufactured goods would be increased.

Bankruptcies continued to take a toll on the ironmasters. In March 1822 William Turton was made bankrupt. He had shares in various ventures including Highfields Ironworks, Highfields Rolling Mill, Tipton Forge, Tipton Green Furnaces and the Moat Colliery. The Turton family had lived in Gornal, but had moved to West Bromwich, where William had been a successful nail ironmonger. Even though William, the father, had now died, creditors pressed both the son (also William) and the estate of Turton and Penn nail ironmongers.

The Turtons were also related to the Fereday family; Jane Turton was married to John Fereday, stonemason. Both John Turton Fereday, ironmaster, and William Turton Fereday, coalmaster, were their sons. J. T. and W. T. Fereday, William Turton and John Blair were also all involved in a partnership that made the River Sow navigable from the Forebridge Wharf at Stafford to the Staffordshire and Worcestester Canal at St Thomas'. They therefore enabled coal to be brought into Stafford by boat.

Brothers John and William Firmstone now controlled the wealthy Highfields mineral estate as well as the Highfields Furnaces, foundry and rolling mills and Roughills Furnaces. The domino effect that was rippling through the iron trade eventually reached the Firmstones during 1821. Both John and William faced the bankruptcy courts in 1821 and 1822 and their estate suffered years of control by trustees until the eventual piecemeal sale. The Highfields Estate was bordered by Capponfield, Barborsfield and Hallfields and was roughly triangular with two groups of ironworks located on the canalside. Highfields Ironworks was let and sub let to a number of ironmasters, but remained in the ownership of the Firmstone family.

Highfields Furnaces were located on the Highfields Colliery and were served by a blast engine with a 42in-diameter cylinder. The furnaces could make 120 tons of iron a week. The plant included a refinery, a cupola, a blacksmith's shop and carpenter's shop. These were advertised for sale in 1824. The other Firmstone properties at Roughills comprised two furnaces that could make between 120 and 140 tons of iron per week and possessed a blast engine with a 56in-diameter cylinder. A third blast furnace was under construction. This estate included coal and ironstone mines and railways connecting the mines with the furnaces. It was advertised for sale in December 1823. John Parsons Firmstone took the brunt of the bankruptcy proceedings while his brothers moved onto other ventures. Some fifteen years would pass before his debts could be settled. Meanwhile, John went into business as an iron merchant. The Parsons family, related to the Firmstones by marriage, faced similar financial difficulties during 1821 which led to the sale of stock at their Barnetts Leasowe Furnaces near Broseley. Within a year Charles Phillips and William Parsons were made bankrupt and James Foster then took charge of the furnaces.

William Firmstone had faced a harrowing experience at the hands of a mob. Colliers invaded the mines in 1822 and roughly handled Mr Firmstone at the Roughills Furnaces; it was an event that even made the Inverness papers in May that year. These were troubled times for the working population who relied on regular employment. Mine and furnace closures, short times and reduced wages were contributory factors in a growing unrest amongst the miners and ironworkers. The unrest came to the surface in April 1822. Coal and ironstone miners had withdrawn their labour through a reduction of wages. For the employers it was a necessary step to save what they could of their businesses. The reduction was considerable, up to a third, and there was the new spectre of the truck system. Out of work miners wandered the countryside begging for money while their families were left to the parish to subsidise their existence. The matter came to a head when striking miners tried to force working miners to cease work. They visited John Turton Fereday's Monmore Green Colliery with the intention of 'ducking' the miners when they came to the surface. The military were there for the protection of the workers and shots were fired. A man named John Robson was fatally wounded and seven other men were arrested for assaulting working colliers. In Monmouthshire, where Midland ironmasters had key interests, the militia were similarly involved to keep striking miners under control. Here, some of the strikers had developed almost guerrilla tactics through dressing in women's clothing and blacking their faces and using these disguises to intimidate working miners.

Rioters who escaped were pursued, but in South Staffordshire some avoided capture. The military adopted a new tactic. In May 1822 Lieutenant Colonel Littleton decided to tackle the chief culprits that caused the rioting at Roughills. At 3 a.m. on a Saturday morning Littleton's Teddesley Troop marched out with a magistrate (the Reverend J. Clare) to Ettingshall Lane where most of the rioters lived. This area was locally known as 'Hell Lane' and had a history of being a home for law-breakers. The timing of their attack coincided with the arrival of Captain Musgrave and the Staffordshire Militia, a new force of special constables, and the Reverend Leigh, a magistrate from Bilston. Houses were surrounded and seven men were committed to Stafford Assizes for punishment.

Meanwhile, the South Staffordshire ironmasters found a champion amongst their numbers, Thomas Price, who argued their cause in an uncompromising manner. He chaired the meeting of coal and ironmasters at the George Inn, Walsall, on Tuesday 7 May 1822. The following resolution was agreed by the principal coal and ironmasters and was to set the policy for the future:

It was unanimously resolved that the sole and entire cause of the disturbances has been a refusal of the workmen to work at the rates of the wages which have been offered them (which exceed those given in other mining districts), and not, as hath been industriously propagated, the payment of wages otherwise than in money.

That the present low prices of all the necessaries of life of life, the reduced prices of coal and iron, render such refusal on the part of the workmen highly injudicious, unreasonable, and unwarrantable, tending to prolong the disturbance of the public peace, to drive the iron and coal trades to other parts of the kingdom. And eventually to produce the ruin and starvation of the workmen and their families; and therefore, that this meeting will neither agree to, or sanction, any other rates of wages than those, which the workmen now refuse to accept.

Aris's Gazette, 13 May 1822

Reinforced by the yeomanry and cavalry, order was restored and miners went back to work at the lower rates, but as will be noted later, Thomas Price paid a heavy penalty for his intervention.

This was a generation that spawned innovation and invention on a grand scale. One who deserves far more credit than historians have given him was William James (1771–1837). James was born in Henley in Arden, the son of a solicitor, William James. The younger James studied the law and was appointed land agent to the Earl of Warwick. He became associated with various mining ventures including mines at Wednesbury which he, trading as the Warwick Mining Co., had leased from the executors of John Wood in 1800.

James owned various limestone quarries and was a promoter of the Upper Avon Navigation, the Stratford-upon-Avon Canal extension from Lapworth to Stratford and the Stratford and Morton Tramway that linked Stratford-upon-Avon with Moreton-in-the-Marsh. William James regularly attended Stratford Canal meetings from 1808 until 1821 and was frequently deputy chairman or chairman. His purchase of the Upper Avon Navigation in 1813 was part of his greater plans for canal and river transport in the region. Once the Stratford Canal was finished in 1816 an important link was established between this canal, the Avon and the River Severn.

William James speculated in various schemes, including an unsuccessful venture where shafts were sunk to find coal at Bexhill in 1806–1809. Fellow coalmaster John Bagnall (1759–1828) was one of the eleven shareholders with William James in the Sussex Mining Co. James lived for a number of years (*c.*1812–1823) at Hill Top, West Bromwich, where he is believed to have owned two houses. He was living there during the miners' riots of 1815–1816 and he put together a petition to Lord Liverpool on the distress of the Black Country miners and ironmasters in 1816.

William was a keen supporter of railways and, in addition to the Stratford & Moreton Tramway, supported the construction of the Liverpool & Manchester Railway. Such diverse interests considerably extended his resources. By 1816 he was encountering financial difficulties and debts went unpaid. His creditors tolerated the situation for some seven years, but finally James was made bankrupt in 1823. It was a harsh blow that destroyed a promising career. His 'administrators' sold the large house in 1824 to help pay his debts and the other was rented for a time by the Attwood family.

Between 1820 and 1822 the firm of Devey, Sanders & Tickell had spent resources on forges and mills at Golds Hill, Leabrook and Toll End. They also operated perhaps the only alum works in the West Midlands, which also was located at Golds Hill. Tickell's bankruptcy caused the closure and breaking-up of the firm. Another iron industry entrepreneur, William Aston, took charge of the Golds Hill Ironworks.

For Aston, his rapid acquisition of property chiefly between 1824 and 1826 was, in hindsight, perhaps ill-advised. He had begun carefully enough with partnerships including those with William Baldwin and Richard Haines. Baldwin's death in May 1824 left Aston in complete charge of the Eagle Furnaces and increased shareholdings in the Horsley Road Collieries and mines at Ocker Hill. Aston's brief time as an iron industry magnate ended with bankruptcy in June 1826. His properties included at this time:

1. Golds Hill Ironworks leased from Thomas Hill.
2. Cop Hall Estate demised to William Aston and Richard Haines by Francis Finch.
3. Highfields Colliery demised by Mr Hartshorne.
4. Eagle Furnaces and Great Bridge Ironworks from Hateley & Hawkes.
5. Meadows Colliery, Sedgley from Hordern and Molineux.
6. Caponfield Ironworks & Colliery from Hordern & Molineux and Bishton & Underhill.
7. Moseley Hole Mineral Estate from Joseph Lane and J.L. Manby.
8. Great Bridge Mines from Dovey Hawkesford and John Hawkesford.
9. Roughills Furnaces, Wolverhampton, from J. & W. Firmstone.
10. Woodcock Street Rolling Mill, Birmingham from Thomas Barrs.

New Investment in the Iron Trade

The iron industry experienced many years in the doldrums, but these were also formative times for establishing new ironworks. It was an occupation for the prudent. Failure in the iron trade became commonplace. Coseley (Deepfields) Ironworks, previously the subject of the bankruptcy proceeding regarding George Stokes, suffered again through creditors trying to recoup their debts. In 1822 the Coseley Furnaces and Ironworks was put up for lease following a request by the High Court of Chancery regarding Thomas and Edwin Pemberton. Their estate then comprised two blast furnaces and rolling and slitting mills. The Pembertons had gained control of the Coseley Ironworks following the failure of George Stokes. Samuel Pemberton (born *c*.1747) had died in August 1803 and his share had passed to his executors: Mary Pemberton (wife), Edwin Pemberton, Thomas Pemberton and Benjamin Stokes.

A partnership reorganisation favoured brothers Edwin and Thomas Pemberton who took charge of the Coseley operations. The Pembertons were a resilient family and were able to overcome the financial problems and retained their connection with the Coseley Furnaces and Ironworks. Following the 1822 bankruptcy proceedings the furnaces and ironworks were worked in the name of Edwin Pemberton (1783–1851).

Samuel Pemberton, the head of the family, had married Mary Grosvenor. He made Birmingham his home and he was engaged in various occupations including that of a merchant and jeweller. He was also briefly in partnership with Robert Samuel Skey, canal carrier and paper merchant. His son, Thomas (1775–1830), continued the jewellery trade and was also associated with clockmaking and the brass trade.

Despite the lengthy depression in trade, new investment remained available. By 1825 some investors were confident enough to start fresh ventures. William Seagar Wheeley and John Wheeley, glass manufacturers of Kingswinsford, leased mines of coal and ironstone from the Brettell family that were placed near, and beside, the Stourbridge Canal at Brettell Lane, and established blast furnaces and ironworks.

The district between Wednesbury and Darlaston attracted investment from ironmasters during the second decade of the nineteenth century. Matthews & Finch and John Lloyd became ironmasters there during this period. Successive generations of the Lloyd family had been involved with colliery operations at Wednesbury and Darlaston ever since Richard Parkes's death in 1729. As one of the heirs of Parkes, the Lloyds controlled a mineral estate of 512 acres. It proved to be a fortunate acquisition as furnaces and ironworks were later established on the estate.

In November 1815 Sampson Lloyd arranged for the Broadwaters Branch that served the canal pumping engine to be deepened and a wharf was made near to the Old Park Colliery. During the next month Samuel Lloyd, acting on behalf of the heirs of Parkes, requested that the Birmingham Canal Navigations grant a long lease for the exclusive occupation of the Broadwaters Branch. It was a request that was not granted because others might wish to use it, especially as the Broadwaters Furnaces were nearing completion at the time. In 1818 John Lloyd & Co. established a blast furnace

An engraving of Corbyns Hall Ironworks when tenanted by Henry Sparrow.

on the Old Park Colliery and by 1823 it was capable of producing 2,660 tons of pig iron a year. In 1828 this production was doubled when a second furnace was put into blast.

Izon & Whitehurst had the Old Level Foundry and Furnaces at Brierley Hill, while Gibbons operated Level Furnaces and set up new mines and furnaces near Corbyns Hall. The development of mines and furnaces in the Corbyns Hall area was pioneered by the Gibbons family, who arranged for the construction of tramway links to the Fens Branch of the Stourbridge Canal. James Foster also arranged for the construction of the Shut End Ironworks and Furnaces which were linked by a railway to the Staffordshire & Worcestershire Canal at the Ashwood Basin. The Firmstone family also moved away from Highfields to set up a new venture near Leys Farm, on the Stourbridge Canal, that became the Leys Furnaces.

Coneygre (or Dudley Port) Furnace was out of work for a period after Zachariah Parkes & Co. gave up the operation. During December 1827 a new lease was agreed between Lord Ward and George Parker of Hill Top. This new operation was distinct from Parker & Co., which continued to operate furnaces at Oldbury and Tipton. The average make of iron was about 45 tons a week. Mines on the adjacent Coneygre Colliery supplied both coal and ironstone to this furnace. Much of the coal raised came from the Thick Coal seam but there was also some Heathen Coal.

Those involved in the ironstone trade were evidently least affected by the slump and were often well placed to finance new smelting operations. John Bagnall, George Jones and Thomas Timmins were numbered amongst the ironstone masters who successfully made the transition.

Thomas Timmins & Co. gradually improved their business ventures in the district. The partnership essentially included Joseph Tarratt and Thomas Timmins, who were shareholders in the Highfields Colliery. Tarratt had also been associated with the short-lived Highfields Furnaces, while Timmins had been a partner with Philip Williams and William Hanbury Sparrow in Pennsfield Colliery. Timmins & Co. developed the Wolverhampton Colliery Estate. These were initially coal and ironstone mines placed along the northern perimeter of the Roughhills Colliery. The company set about the erection of a blast furnace there during 1825 and subsequently added two more. The adjacent Roughills Furnaces, which had passed to William Aston ceased to be used and were eventually pulled down. Joseph Tarratt remained a partner in Wolverhampton Furnaces until 1835.

John Bagnall & Sons, Ironmasters

The firm of John Bagnall & Sons built up their business over a period of time through sensible acquisitions. Accounts written by both Frederick Hackwood and Samuel Griffiths have aided historians in their understanding of how the firm of John Bagnall & Sons was established, although some of the reported facts deserve closer scrutiny. Samuel Griffith, in writing his book on the iron trade provided good illustrations and a lot of useful facts, but also included information of a less sound basis. His account of Richard Foley and the method of slitting iron he obtained from a Russian ironworks were proved to be more fiction than fact. It is also possible that the reference to the death of Edward Bagnall, brother to John, deserves further attention. Griffith regarded the death of Edward Bagnall as a crucial stage in the history of the firm because it gave John control of the business. The unfortunate problem with families closely involved in the same trade is that names commonly reappear from generation to generation; there were at least three Edward Bagnalls associated with the local coal and iron trades.

The family links begin at Broseley, Shropshire, where John Bagnall married Margaret Dixon in 1753 at St Leonard's church. He is known to have moved to the Darlaston area where he died in 1800. His job has been quoted variously as surveyor, bailiff or butty. Connections have been made between this John Bagnall and John Wilkinson, ironmaster, who set up the Bradley Ironworks near Darlaston and would have needed men and managers to run his Bradley mineral operations. John Wilkinson was a resident of Broseley at this period and a working association between the two men cannot be ignored.

John Bagnall had three sons baptised at St Leonard's church, Broseley: William (29 September 1754), John (15 July 1759) and Edward (22 March 1761). There was also Daniel, born in 1765 after the time when John appears to have left Broseley. All were seemingly engaged in the coal trade initially as miners. There was also a strong religious bond that united the family: they were Wesleyan Methodists and as such frequented the various chapels or attended the circuit services. Records of their respective families start to be recorded in the registers of Darlington Street Wesleyan chapel at Wolverhampton, Willenhall chapel, the Wednesbury and Darlaston Circuit and King Street chapel in Dudley from the 1790s.

Edward Bagnall, brother of John, died in 1805. According to Griffith's *Guide to the Iron Trade*, Edward and John had been partners in colliery ventures at Coppice, Leabrook and Toll End, as well

A postcard view of Leabrook Ironworks. These works were constructed for and owned by John Bagnall & Sons. Although close to Wednesbury, Leabrook was part of Tipton and had the distinction of having four ironworks and a blast furnace named after this place.

as at the Leabrook Furnaces, near Wednesbury. A check of local archive records reveal that little of this statement was true. Leabrook Ironworks had actually been established by Michael Toney, a Wednesbury victualler. These works specialised in providing parts for the gun trade. Wednesbury was then an important centre for gun making, where a number of families were engaged in gun-lock forging and filing. During 1803 Toney advertised the Leabrook Ironworks for sale and it was this year that has been taken as the establishment of these ironworks. Partnership records as published in the *London Gazette* show that Thomas Bolton was a partner of Michael Toney at Leabrook until 1809 when Toney left the partnership. There was then another brief association of Bolton and Isabella Pardew that ended in 1810. The furnaces and colliery seem to have been an independent venture. There is no mention of Leabrook Furnace in a list of British blast furnaces compiled in 1806, but one is shown on Sherriff's 1812 map of the mining district. Leabrook Furnace and Foundry was advertised for sale in 1811 following the bankruptcy of John Read. So it would seem that Leabrook Furnace was briefly part of Read's empire, and perhaps established by or for him.

The first traceable reference connecting the name of Bagnall with Leabrook Furnace was the 1818 Staffordshire *Directory* that lists under 'Bilston', John Bagnall & Co. as at Toll End and Leabrook and under 'Wednesbury' as ironmasters, Tipton. This firm held the furnaces until about 1823. In April 1823 Leabrook Furnaces were available for lease. The property then included a blast furnace with fineries, foundry, pattern shops and weighing machines. There was a new hearth and the furnace was in good repair. The Birmingham *Trade Directory* of 1823 records Leabrook Furnace as being in the ownership of the Leabrook Furnace Co. This latter company may have been a partnership of Thomas Matthews, William Davis and Joseph Warr who were made bankrupt in 1822 and were described as ironmasters of Leabrook.

As for John Bagnall & Co., this partnership was created by 1806 when John Bagnall, coalmaster of Wednesbury, William Turton, coalmaster, and Charles Norton, lime merchant of Birmingham, leased an extensive tract of land at Toll End from Thomas Bannister. This property was bounded by the Tame, Forge Pools and the canal. In 1808 further property, known as the 'Wednesbury New Line Collieries', was also leased from Thomas Bannister. During the intervening two years, John Bagnall had moved home from Wednesbury to live in West Bromwich.

The principal line of the Bagnall family as ironmasters came from John, the son of John Bagnall of Broseley. This John had been married twice. His first wife, according to Hackwood, is said to have given him a son, Edward, and two daughters. Darlaston parish records identify the daughters as Nancy and Phoebe. There was also an Isaac, who died young. Phoebe, the mother, died in 1791. John Bagnall married again. John Bagnall (of 1759) married Mary Royal at Wednesbury in 1792 and had a number of male heirs by this union: John (born 1794), William (born 1797), Thomas (born 1799), Henry (born 1800), Samuel (born 1803), James (born 1804), George (born 1806) and Richard (born 1809). His partnership in ventures such as Leabrook Furnaces and Leabrook Colliery was as John Bagnall & Co., but with bankruptcy proceedings against both Charles Norton and William Turton, John acquired the controlling interest. If he was involved with the Leabrook Ironworks, this association was evidently over by 1819. For it was in December of that year when Michael Toney decided to grant a lease to Henry Devey, Jonathan Sanders and Thomas Tickell for the forge mill and engine.

Edward Bagnall (of 1761) married Mary Wilkes at Darlaston in January 1783 and continued to reside in the Wednesbury and Darlaston district, building up a trade as a coalmaster. Edward Bagnall was also buried at Darlaston. When Edward died in December 1805, he left a will which made provision for his wife Mary and his children, including sons Daniel and Edward. This will was proved at Lichfield in June 1806 and appointed his brother, John, and Thomas Bayley as executors. The executors were to continue to operate the mines and from the profits raise a capital sum of £1,000 that was to be invested to provide money for Mary and her children. When, and if, the minerals were exhausted they were also to sell off horses, engines and plant and the revenue to be divided between Edward Bagnall's family. There was no mention in the will of any ironwork ownership.

Edward senior's colliery ventures may well have included the Millpool Colliery. The *London Gazette* recorded the partnership change of Daniel Bagnall senior, Daniel Bagnall junior, Edward Bagnall and Jeremiah Dimmack in the Millpool Colliery near Wednesbury – Daniel senior left this partnership in 1810. Later in 1817 Edward Bagnall was mentioned in Birmingham Canal Navigation minutes as the owner of the Millpool Colliery engine when an offer was made to supply water to the canal company. The Staffordshire *Directory* for 1818, by comparison, lists John Bagnall as the owner of Millpool Colliery. The Daniel (junior) and Edward mentioned might have been the sons of Edward (1761). The other Daniel was probably their uncle, and brother to John. This Daniel appears to be the coalmaster, Daniel Bagnall, who was mentioned in trust deed of the Independent chapel at King Street, in 1820. A Daniel Bagnall was also agent for the Coltham Furnaces and Collieries during the mid-1830s.

The names of John and Edward Bagnall were finally linked in a proposal to the Birmingham Canal in 1823 to make a basin for their Tividale Colliery. Tividale was a completely different venture and partnership. These mines were to be found alongside the Old Main Canal on the boundary between Oldbury and Tipton. The Edward, in this case, might have been the step-brother of John (1794). In later years the partnership for Tividale Colliery would include James Bagnall, Richard Bagnall, Thomas Bagnall, William Bagnall, Thomas Jesson and Henry Smith. The development of Tividale coincided with the closure of Leabrook Colliery. This mine, by 1823, had been worked out. The mine pumping engine, whimseys, gins, wagons and rails used at the pit were advertised for sale and interested parties were asked to contact John Bagnall at Hill Top.

Seeds of Change for Transport Improvements

The 1820s may have been a time of industrial depression, but they were also a time when public railways became a reality with the construction of the Liverpool & Manchester, Stockton & Darlington and the local line from Stratford-upon-Avon to Moreton-in-the-Marsh. But, these schemes were small compared to the national canal network that provided the principal means for the ironmasters to move their goods. Local waterways profited by the tolls on goods that passed along their respective routes. Ironmasters were particularly keen to challenge these rates and sometimes achieved drawbacks on specific traffic. Ironmasters did have a choice in transport; there was also a network of turnpikes, and carriers operated stage wagons over lengthy distances. In some cases, road transport was the only means of carriage. The Sheffield trade particular relied on wagons using the turnpikes and roads through the East Midlands to South Yorkshire. Road transport was equally, if not more, expensive as tolls were charged by each turnpike trust.

Several railway and long-distance tramroads were suggested to break the exclusive control of the canals and turnpikes. William Mathews was one individual who was particularly incensed with the charges made by the Staffordshire & Worcester Canal Co. and was part of a delegation who appproached the canal company with regards to tolls made on traffic passing to Liverpool and other parts in the North West. William later recounted, in evidence to the parliamentary Select Committee on Railway and Canal Bills (1853), that their applications were disregarded and upon threatening to build a railway to Liverpool were laughed at. The deputation then adjourned to a hotel in Wolverhampton where plans to construct a railway from Birmingham to Liverpool were fist formulated. Although this scheme failed there was a gathering mood for railway construction.

Some canal companies were not insensitive to this mood and made improvements to their systems to speed up trade. The Birmingham Canal Navigations employed Thomas Telford to engineer a new and shorter route between Birmingham and Wolverhampton. The Oxford Canal made significant alterations and shortenings to their line between Hawkesbury and Braunston, while a brand new canal was made to Telford's designs, linking Nantwich in Cheshire and Wolverhampton and providing an alternative and more direct route to Liverpool. The seeds for railway competition were sown, however, and within ten years an embryonic railway network was established.

CHAPTER 4

THE WILDERNESS YEARS
TRADE DEPRESSION AND CAUTIOUS REVIVAL,
1827–1837

The percentage of working blast furnaces remained high in spite of the prevailing depression. A list of furnaces in the Staffordshire and Worcestershire districts for 1826 mentions a total of 109 furnaces with 90 in blast making nearly 200,000 tons of pig iron.

Investment in new ironworks was conducted cautiously and often sensibly. Investors and ironmasters made some prudent decisions that were to set up a group of successful and long-lived operations. Their success was often down to sheer business skills, but was also aided by the organisation created by the ironmasters themselves. Through their quarterly meeting they were able to regulate prices and co-operate with each other, to a degree, at a very difficult time for the local iron trade.

The Bagnalls built up mineral estates in the area of the River Tame where the parishes of Tipton, Wednesbury and West Bromwich met. It was at nearby Hill Top that members of the Bagnall family made their home. John Bagnall decided to take five of his sons, by his second marriage, into partnership. They were John, William, Samuel, Thomas and James. All took an active part in running the firm, which then became known as 'John Bagnall & Sons' in 1828. John Bagnall established a reputation for good quality iron. Their Crown I.B. brand was a popular choice with firms working iron up into finished items. John Bagnall & Sons were set on a course of business expansion. Goldsgreen Furnaces were commenced in 1820. Two furnaces were built and a third was added in 1825. A long basin served the furnaces that joined the Balls Hill Branch. Here minerals would be unloaded for the furnaces and pig iron despatched to works served by the Birmingham Level. They also built a new ironworks beside the Danks Branch, which joined with the Walsall Canal below Ryders Green Locks. A tramroad connected Goldsgreen Ironworks with the furnaces and it likely that minerals and pigs were sent along this tramway.

Golds Hill Ironworks was placed close to the Tame and was mentioned in various deeds as the 'water-powered Golds Hill Forge', known sometimes as 'New Goulds Hill Forge'. This was added to the Bagnall estates as part of the properties accumulated from William Aston after his bankruptcy of 1826. The forge had a variety of owners including Joseph and John Booth Hodgetts of Dudley, and from 1815 the owner of Blanaevon Ironworks, Thomas Hill. Operators also varied. John Read, ironmaster of Gospel Oak, had the forge until his bankruptcy. Later Devey, Tickell and Sanders had the forge, then William Aston and finally Bagnall & Sons. The agreement for the lease of Golds Hill Ironworks was made between Mary Hill, widow of Thomas Hill, and John Bagnall & Sons in 1827. These ironworks were placed at the terminus of the Danks Branch

There was another riverside mill that was also called 'Golds Hill', which lay to the south of Hill's mill. Trade directories first mention a flour mill there in the ownership of Davis and Bagnall, while the Tipton Rates for 1836 list Davis and Bagnall having coal pits at nearby Toll End. The Davis mentioned was Thomas Davis, a Wesleyan timber merchant, who had married Phoebe Bagnall. An ironworks was erected beside the flour mill which was worked by the partnership of Davis & Bloomer. In 1857 they were recorded as owning eleven puddling furnaces. These works were also placed beside the Danks Branch, also close to the Tame Valley Canal. Golds Hill corn mill appeared regularly in directories from 1834–1856.

John Bagnall & Sons was effectively a partnership of John Bagnall (1759) and sons James, John, Samuel, Thomas and William, with Richard involved in certain colliery partnerships. His other sons became involved in other disciplines. George went to live in Cheltenham where he practiced as a doctor. Henry went to Cambridge University in 1820, where he was admitted to Queen's College where he trained for Holy Orders and was ordained priest in 1825. He was at Lichfield, Shieston (in Shropshire), and finally vicar of Great Barr.

The Bagnall brothers gained complete control of the firm from 1828, when age finally caught up with their father. He effectively retired from the business and died the next year at West Bromwich. The sons now started to make considerable investments in new property.

William Aston's bankruptcy led to a number of works coming onto the market that included the Caponfield Furnaces, which were acquired by the Bagnalls. Caponfield Furnaces were located beside the Old Birmingham Canal at Bilston; they were somewhat isolated from the other works but soon became an integral part of the Bagnall empire and frequently despatched iron to their ironworks along the banks of the BCN. Caponfield Furnaces once belonged to the partnership of Smith and Read. By 1820 this property included two blast furnaces, a casting house, fineries, grinding blade mills and a blacksmith's shop that was occupied by Thomas Smith and his under tenants. A half-share in this restate was offered for sale in 1820, but was not enough to prevent the bankruptcy of Thomas Smith in 1821. William Aston eventually gained control of the furnaces and associated mines of coal and ironstone. These were again advertised for sale in January 1827 following Aston's own bankruptcy. There was also a freestone quarry worked by Flavill and Grier.

Mining properties were a frequently changing asset because as some mines became exhausted others were developed. New coal and ironstone mines were frequently opened to maintain the supply to the furnaces and ironworks. In 1831 at Dunkirk Hall, a short canal branch was made to serve the the mines, linking with the Birmingham Canal New Main Line. The Crescent Colliery, near Willenhall, was worked, which required a horse-worked tramway across Gypsy Lane to the canal. James Bridge's coal and ironstone mines were leased from the Earl of Bradford once the Price family and their executors ceased to work them, while other mines were developed at Bentley Deepmoor, near Walsall. Here tramways fanned out to various pits from a central wharf on the Ansons Branch Canal. Further ironworks were acquired which included the Toll End Ironworks (formerly Woolley & Pretty) and the Imperial Ironworks at Wednesbury. All became part of an integrated business of mining and ironworking that was to ensure orders from railway companies and continued prosperity for the company.

Meanwhile, a separate business course was followed by Edward Bagnall. The story of Edward Bagnall, the son of John by his first marriage, is most complex. It is a name that crops up from time to time in the 1820s and 1830s, variously as coalmaster, mine agent, coal merchant, ironmaster and land agent. From piecing together the different references made to Edward Bagnall, the following may apply to him. It seems he lived for a time in Birmingham, acting as mine agent. Following the death of John Bagnall it is probable that Edward Bagnall received a share of the estate; his name became linked with mines at Toll End and the Wednesbury Line Collieries. He had sufficient capital and was able to invest in railway companies such as the Birmingham, Wolverhampton & Stour Valley in 1845. The 1841 census finds him living with brother-in-law, Thomas Davis, a timber merchant at Goldsgreen, who married his sister Phoebe. This Edward was associated with Davis in coalmines at Toll End and the Golds Hill flour mill. When the 1851 census was taken Edward was still living at the home of Thomas Davis, timber merchant of Golds Hill. His status as brother-in-law to Thomas was confirmed and his age was given as sixty-nine, which puts his year of birth as either 1781 or 1782. Searches have yet to find his place of birth, baptism details or the marriage of his mother, Phoebe, and John Bagnall. Edward died unmarried in 1859. Caleb Bloomer was an executor of his will.

The Birmingham Coal Co. was a neighbour of the Bagnalls. The firm was a joint stock company which had been formed in 1793. A group of Birmingham businessmen concerned about the price of coal decided become joint traders in the business of purchasing, working and vending mines of coal:

Articles for Birmingham Coal Company – Formed March 25th 1793

Whereas from various incidents and circumstances, coals in Birmingham have, of late, not only very much advanced in price, but are become very scarce; and if some methods are not adopted to remedy such evil, very serious injuries may result therefrom to the Manufactories of the Town. AND WHEREAS, it appears to all and every Parties hereto, that nothing can effectually prevent the Price of Coals from increasing, and tend to make that necessary Article more plentiful than by forming themselves into a Company or Copartnership for the purpose of purchasing or working Mines of Coal, an vending same in Birmingham aforesaid, or purchasing and vending coals only, as may hereafter be thought most conducive to the ends above mentioned

Birmingham Reference Library 6506

This agreement eventually led to the working of coal and ironstone mines near Hill Top on the Old Wednesbury Canal, at Toll End and at Darlaston Green. By 1812 a furnace had been erected on the Toll End Estate, close to the Walsall Canal. The Birmingham Coal Co. eventually had three furnaces at Toll End which remained in work through to the 1830s. There was also an ironworks on the site, also known as 'Toll End'. Between 1837 and 1841 attempts were made to sell off the furnaces at Toll End and their cement works at Darlaston Green.

Thomas and Henry Price assembled an integrated network of mines and works. During June 1823 they signed a tonnage agreement with the Birmingham Canal Navigations Co. for the payment of a set annual rate of £450 for their conveyance of raw material used in their respective works. These were specified as mines at Glebefields, Schoolfield, Groveland, Princes End, Daisey Bank and James Bridge, the Walsall Limeworks, blast furnaces at Bilston Brook and Glebefield and their West Bromwich Forge and Mill. Their Glebefield Furnace had a relatively short existence. From the name, the location would appear to have been at the Glebefield Colliery that was placed to the south of J.T. Fereday's furnaces at Bradley.

The Price family lived for a time at Charlemont Hall. Mary Willetts, in her *History of West Bromwich* (1882), described the Prices who were living there in 1822. She specifically noted their 'sad history' and alleged that they were said to have taken the roof from Tipton Old Church to cover an engine house and 'no good came after that'. No mention was made in Willetts' account of the prestige the elder Thomas Price had gained amongst his fellow ironmasters. Nor does she mention the respect he gained of others. The agent for the Earl of Bradford was often quite complimentary about Thomas Price in correspondence he had with the Earl.

'Sad history' may be an understatement when describing the fortunes of the Price family. 'Cursed' may well be a better term! Their tombstone at the south end of All Saints' churchyard, West Bromwich, testified to the comparatively early deaths of the children. One of the sons returned home intoxicated one night and, being unable to obtain an entrance into the house, lay on the lawn the whole night. He took cold and died soon afterwards. Mr Price was found dead with his throat cut in one of the bedrooms – although the fact was kept quiet and no inquest held, according to Mrs Willetts there was little doubt of its correctness.

Thomas Price married Mary de Bois at Sedgley in June 1791. They had five children, Henry Lemm, Mary Ann, Richard, Thomas Blois and William George. All were buried at All Saints', West Bromwich. The catalogue of deaths began with Mary Ann, who died in April 1820. Thomas Price senior died on 4 December 1822, William George died on 19 December and Richard on 1 June 1825. Thomas and Henry continued the business for another eleven years. Thomas Blois Price finally passed away on 6 February 1836 after a long illness. Henry Lemm Price died on 2 April 1836, leaving the business in the hands of the executors, W. & J. Millington of Summerhill Ironworks, Tibbington.

Aris's Gazette printed the announcement of Thomas Price's, demise. It was described as sudden and lamentable. If Mary Willetts's account is to be believed, his death was also murder. His role as champion of the ironmasters may well have created a few enemies within the mining community. It was a time when the fledgling Ironmasters Organisation was gaining strength and importance.

They continued to meet every quarter-day, however, and undeterred by the loss went onto form a body that set prices and became an important force in the local iron trade.

The Sparrow Brothers made some important business investments that laid the foundations for a strong ironmaking and ironworking concern. In 1812 William Hanbury Sparrow and William Hanbury had leased the Old Brownhills Colliery from Phineas Hussey; it was to be an important supplier of coal for the Bilston Ironworks and Furnaces. Between 1822 and 1823 an 120-acre estate at Stow Heath was leased. This extensive mineral property contained mines of coal and ironstone that William Hanbury Sparrow and John Sly Sparrow exploited with the construction of furnaces. The first Stow Heath blast furnace went into blast in 1824. There were five furnaces on this site eventually. Tramways linked the mines and furnaces with the Bilston Mill Ironworks.

John Sly Sparrow died prematurely in 1825, leaving a young wife, Charlotte (Penn), whom he had married in 1817, and a young son, John William, born in 1820. Charlotte retained an interest in the Stow Heath and Bilston estates as trustee until her son was of an age to take his father's place. In June 1844 Charlotte relinquished her share in the firm of W. & J.S. Sparrow and John William Sparrow took over his mother's interests.

William Hanbury Sparrow increased his separate ironworks holdings with the Lane End Ironworks near Longton in North Staffordshire. These works were established in the mid-1820s to work local seams of ironstone. The Trent & Mersey Canal tramway from Stoke to Lane End provided the link for moving their iron from Lane End to Stoke and from there carriers could ship it along the canal network.

The Parker family chose this period to withdraw from ironmaking at both Oldbury and Tipton after 1830. The original Tipton Furnace had closed and their other works near Bloomfield was transferred to Edward Cresswell, while Oldbury passed to the Dawes family. George Parker took on a new challenge, that of working Coneygree Furnace. An agreement was made on 29 December 1826 between Francis Downing, on behalf of Viscount Dudley and Ward, and George Parker of Hill Top in the parish of West Bromwich, ironmaster, whereby Francis Downing agreed to let to George Parker the iron furnace at Coneygree in the parish of Tipton together with the blast engine, blowing machinery, finery, iron foundry buildings, shops and dwelling houses belonging to Lord Dudley and Ward at the yearly rent of £330 payable half-yearly. The tenancy was to commence on Lady Day 1827.

Lord Dudley and Ward was to put buildings in good repair and condition, while Mr Parker was to provide a new furnace hearth, to contract and engage to keep iron furnaces, blast engine, blowing machinery, foundry buildings, shops and dwelling houses in repair. Mr Parker also agreed to purchase stocks of tools implements and effects that had formerly been used by the previous tenant, Zachary Parkes & Co. In August 1827 terms were also proposed for George Parker to take over a coal pit then worked by John Hopkins to bring coals to Coneygree Furnace.

James Foster (1786–1853) had a magic touch when it came to the iron trade. He had an early introduction to the trade through the partnership in John Bradley & Co. James and brother, William, were underage when that partnership was first formed in 1802, but gained responsibility with age. James eventually became senior partner in the firm, when John Bradley died in 1826. There were several different partnerships that involved James Foster, including the iron founding and engineering firm of Foster, Raistrick & Co. and the firm of Thomas Jukes Collier & Co., ironmasters of Wombridge and Hadley in Shropshire. But the core business was that of John Bradley & Co., which James remained associated with until the end of his life.

John Urpeth Raistrick (1780–1856) was born in Morpeth, Northumberland, but moved south to gain experience as an engineer and millwright. He was employed at the Ketley Works, near Wellington, and was in partnership with Mr Hazeldine at Bridgenorth Foundry. By 1816 Raistrick was in practice as a civil engineer in his own right. When Hazeldine died in 1817, Raistrick joined the partnership of John Bradley & Co. and set up the business with James Foster to make rolling mills and steam engines. His association with John Bradley & Co. and Foster, Raistrick & Co. lasted until 1831.

James Foster (of Stourbridge) and Henry Bradley (of Wollescote and the only son of John Bradley) then formed a new partnership that continued under the name of John Bradley & Co. as owners and operators of the Stourbridge Old Ironworks, the Stourbridge New Ironworks, Stourbridge Foundry, Shut End Colliery, Scotts Green Colliery, Brierley Ironworks, Baptiste End Colliery and the Shropshire Ironworks at Eardington and Hampton Loade. James Foster also continued the separate Shropshire ventures at Barnetts Leasowe and Wombridge, as well as establishing other iron furnaces at Madeley Court. Such were the profits from these ventures that James Foster moved from Stourbridge to Stourton Castle, the former home of the Homfrays and Michael Grazebrook. Foster made several important alterations to the castle, which remained his home until his death. Stourton Castle lay beside the River Stour and the Stourbridge to Bridgenorth turnpike, which proved convenient for Foster during his brief term of office as MP for Bridgnorth.

The Firmstone family made a new start at the Lays near Brierley Hill when the Lays Farm Estate became available during 1824. A pair of new furnaces were built close to the Stourbridge Canal for George and William Firmstone. Both George and William made their homes in the district. William lived at the Mount, Wordsley. He was the eldest of the Firmstone brothers and had married Sarah Onions from Dawley, in 1810. Their eldest child was Henry Onions Firmstone, who took over from his father as a partner with his uncle George at the Lays, after William died.

Thomas Firmstone was responsible for an ironworks and mining venture at Madeley, Staffordshire. The opening of the Grand Junction Railway during 1837 provided a new transport route through the district. Thomas capitalised upon the opportunity through the lease of land from Lord Crewe in 1838. He set about sinking shafts and arranged for a private railway to link his coal and ironstone mines at Madeley Heath with the public railway. Construction of ironworks and two blast furnaces followed in around 1842.

An important change of ownership was associated with the Attwood partnership at Corngreaves, Dudley Wood and Netherton, which was sold in 1825 to James Henry Shears, Robert Small and John Taylor, who also controlled furnaces and ironworks at Abersychan, near Pontypool, in Monmouthshire. These partners set up the British Iron Co. on 28 April 1825. Abersychan had four blast furnaces, another two were built to make six, to complement the six in the West Midlands. The investment did not end there, for in 1825 the partners purchased the Acrefair Furnaces and Ironworks near Ruabon and Newbridge Ironworks near the River Dee.

Acrefair belonged to an extensive ironworking estate built up by Edward Rowland of Gardden Lodge, Ruabon. Rowland died in 1815 and the properties were inherited by his son, Edward Lloyd Rowland, who left the management to others. An agreement was made with George Homfray (1778–1848), the son of Jeston Homfray, for the venture then known as the 'Ruabon Iron Company'. Homfray wanted to develop the mineral estate, but encountered opposition from Rowland. George was forced to stop the ironworks in 1822, chiefly through funding problems, but also as result of Rowland's own financial affairs. Rowland was made bankrupt in 1823. Acrefair Ironworks, including two blast furnaces and sixteen puddling furnaces were advertised for sale in 1825 and were purchased by the British Iron Co. for £72,000. Newbridge Ironworks was purchased from the Greenhow family for £18,500.

The British Iron Co. looked to rival the Foleys and Feredays for the extent of their ironworking enterprise, but their tenure was dogged with legal disputes with the Attwoods which handicapped what might otherwise have been a profitable venture. The core dispute was that the Attwoods had made false returns regarding the property. The British Iron Co. decided not to make certain payments. Proceedings were taken out by the Attwoods. The case was heard twice before Stafford assizes, went through various courts, including the Court of Chancery, and was finally settled by the House of Lords in favour of the Attwoods. The extensive litigation effectively ruined the British Iron Co. The legal proceedings carried on for years and might have had a different result were it not for the political and banking connections of the Attwood family. During 1843 the British Iron Co. decided to sell their operation to William Routh, Arthur Wilkinson and James Mc Killop for £20,0000. On 2 November 1843 it became the New British Iron Co.

Despite the long-term iron trade depression, new blast furnaces continued to be built beside the Birmingham, Dudley, Stourbridge and Wyrley & Essington Canals. By 1826 there were fifty-one locations listed in South Staffordshire and East Worcestershire. There were 105 blast furnaces and 88 were at work. The average production of pig iron was estimated at 3,704 tons per week. The design of blast furnaces had altered little during the early period. The basic shape was essentially square, with each furnace being made of bricks or stone. Improvements had been made to the blast and capacity, but the make rarely exceeded 40 tons of iron per week. The form and structure also began to change. Furnaces were being constructed conically and round, with the shape held together by metal bands or hoops.

Most of the iron was run off into moulds created in sand, by a process that changed little through the years. A description of the method was published after a visit of the British Association to the Corngreaves Furnaces in 1886:

> The furnace is tapped by some workmen whereupon a thick stream of molten iron immediately commences to run down the side trenches and thence into the moulds at the farthest end of the bed. These being at a lower level than the furnace. As the lower moulds become sufficiently filled a workman bebars the further ingress of metal by closing the trench approach to the mould. He does this by hoeing or scraping up a quantity of sand. The iron then finds its way into higher sections of the bed until a series of moulds is filled.

Bilston Ironmasters

Perhaps the largest concentrations of blast furnaces were to be found in the Bilston district. Furnaces and ironworks were built on all sides of the town. The Wolverhampton Level (473ft) of the Birmingham Canal crossed the district in a wide, meandering loop that headed south from Wolverhampton, passed through a corner of Sedgley and into Bilston Parish where the line of iron furnaces, mines and ironworks became almost constant. The first of the line was Millfields Furnaces and then there were the Bilston Furnaces and Ironworks. The canal then turned eastward and passed back into Sedgley where Coseley, Deepfields, Caponfield and the Highfields Ironworks were located. Just beyond Highfield the canal again passed into Bilston and served Barborsfield Furnaces (erected 1826), Bovereaux Ironworks, Pothouse Bridge Foundry and the former Wilkinson group at Bradley and Hallfield. The canal turned south again through Sedgley to Wednesbury Oak and into Tipton Parish for Gospel Oak and Tibbington where another westward turn brought the course back to Bloomfield, passing Tipton Green Furnaces and Parker's Tipton Furnaces en route. Another, separate waterway terminated on the eastern side of Bilston. Called the 'Bilston Canal', this line was a branch off the Walsall Canal at the 406ft level. This branch served the Bilston Brook Furnaces (belonging to the Price family) as well as a number of collieries and ironstone mines.

The district around Bilston was home to many families involved with the iron and coal trades, who settled there from different parts of the country attracted by the prospect of employment. Their lives were tied somewhat to the successes and failures of their employers and were also subject to their rules. Those whose lives were intertwined with the industry here were also subject to the harshness of employment. It was from Bilston that a concerned member of that community wrote to the *Birmingham Journal* in October 1829 to highlight the truck system that pervaded employment there and elsewhere throughout the district. By this date the payment of wages other than by cash was illegal. The detail, however, was not a deterrent for those intent on avoiding it, just the same as certain employers in modern Britain avoid the minimum wage with immigrant labour.

The letter published in the *Journal* was written by 'Observer' of Bilston and led to a lengthy set of correspondence, some supporting and some disagreeing with the truck system. 'Observer' hinted at extortion and slavery at the hands of the ironmasters. While it was indicated that some masters did not adhere to the truck system, for those who did increased charges on goods varied from a moderate to substantial profit. A picture was painted of women and children waiting hour after hour at the door of the 'Tommy Shop' for their turn to be served, while honest tradesmen lost customers.

As a supplement to this letter the newspaper published the following explanation:

> In order to evade the law, which forbids the payment of wages, other than in cash, some Manufacturers (ironmasters &c) adopt the following trick. They pay the man his money in one room of the office, and then in the next, through which he is obliged to return, is the clerk of the Tommy Shop, who makes his demands, which after the reckoning has been thrown off four, five or more weeks, as is frequently done, in order to force off goods, takes the whole or nearly of what he has received. This game is very common. There is, moreover, a very convenient species of circulating medium in use with some Tommy masters, the adept 'tommy notes' – a workman in want of a few shillings, or some goods, between these long reckonings, goes to the Clerk of works, who instead of money, gives him a scrap of paper bearing these or some words 'let A.B. have goods to the amount of – s-d- (Signed) C.D'. These 'Tommy Notes', as they are emphatically called have in some instances, a considerable local circulation. The man, instead of taking immediately to the Tommy Shop, frequently takes them to the public house for ale, or to the shoe mender (the Tommy Shop provides him with new Shoes), or to the Landlord for rent &c &c, where they are current at 15-20 per cent discount. The person in this way collected a number, sends them to the Tommy Shop by wholesale, and thus many supply, in a great measure, their houses with necessaries. In other cases, the goods are taken and resold at about the above loss, and in this way. The tailor, shoemaker, publican &c &c in the neighbourhood more particularly referred to, is stocked with sugars, soap &c &c often to excess.

Bilston was also a particularly unlovely place to live or work. The sulphurous fumes from the furnaces and ironworks would have pervaded everywhere. Living conditions were such that the town was adversely affected by the Cholera epidemic that swept Britain in the years 1831 and 1832. Dr John Connolly penned a graphic description of the town following his visit in 1832 for the *Warwick & Warwickshire Advertiser*:

> My readers of your paper are probably better acquainted with the mining country of Staffordshire than the writer of this letter can pretend to be. To those who are not it may be sufficient to say that Bilston is situated 11 miles from Birmingham in the midst of a district of which only a small portion of the surface are cultivated – the wealth of the county lying beneath it in extensive beds of iron and coal. The works necessary for availing the proprietors of this wealth are thickly scattered over the district and furnish employment for a large and very crowded population. The singular appearance of numerous fires which belong to the iron furnaces excite the admiration of those who travel through this peculiar community at night; and the volumes of smoke in which this whole county is enveloped by day is hardly remarkable with this smoky atmosphere and with these great fires burning day and night. Bilston is nearly surrounded, and a peculiar region may be said to be produced, where streams of air of very different temperatures rapidly succeed each other. The streets are irregularly and badly paved – the town abounds in courts and alleys – and the whole surface is so unequal that pools of water near the most of the poorest habitations and with the numerous manure heaps and other nuisances would make it impossible for the passenger to pick his way along the lanes and alleys by night without safety, but for the lurid light cast up the most intricate and melancholy passages by the flames of the distant and enormous fires.

Even in later years Bilston remained heavily industrialised. The canalside factories often took water from the canal for cooling purposes and returned it hot. It was not uncommon for mists to cover the canal, especially at Bradley where the works were most concentrated.

Ironstone Resources

Iron produced from the smelting process was extracted from both iron ore and ironstone. The former term is reserved for minerals that have a high content of iron (50 per cent or more), while the latter refers to minerals with a lower content (20-50 per cent). Iron ores include haematite, which was found in different parts of the country. Barrow-in-Furness was once noted for its haematite mines. In the West Midlands lower-grade ironstone was more commonly found. This Blackband ironstone had local names that depended on the measure from which it was taken. The ironstone existed in several strata above and below the Ten-Yard Coal. It was sometimes found in continuous beds known as 'Blue Flats', 'Blue-Clist' and 'Whitestone', but more usually occurred in detached masses of a few pounds that were called 'Balls' or 'Cakes'. These masses had many, and sometimes fanciful, names that include 'Binds', 'Penny-Earth', 'Gubbin Stone' and 'Poor Robin'.

Ironstone was not uniformly distributed throughout the South Staffordshire Coalfield. Some districts provided a good yield, while others were bad. In fact, the mining district can be divided into three major productive areas. The first lay between Brierley Hill, Dudley and Netherton and provided ironstone for furnaces served by the Dudley Canal. The second, and largest, area comprised a crescent-shaped section, which extended from Tipton through Coseley, Ettingshall, Wednesfield, Willenhall, Bilston and Darlaston. The Old Main Line of the Birmingham Canal, the Walsall Canal and Wyrley & Essington Canal all reached parts of this district. The third area extended northwards from Walsall through Bloxwich to Pelsall and was served by the Wyrley & Essington Canal.

The stone had been taken from South Staffordshire in increasing amounts for more than 300 years. Parts of the exposed mineral strata supplied stone to early water-powered furnaces such as Rushall, Grange and Cradley during the first decades of the eighteenth century. Later shafts were sunk and the stone was mined like coal. Ironstone miners usually worked in much more confined conditions than the coal miner. Each worker stooped or crawled to the face though a heading that was, at best, 3ft high. For many local coal miners the Thick Coal often provided cavernous working spaces. In an ironstone mine a head, or passage, was driven to the furthest extent. The miners then worked back and loosened the mass with their picks. The ironstone was extracted and the waste pushed back by a process that was known as 'gobbing up'. Timbers supported the roof and were removed after bed was exhausted. The old passage was then allowed to fill naturally with the settling of the ground.

Local ironstone had provided the impetus for establishing the industry and the development was down to a small nucleus of families who set up the coal and iron trade. Successive generations carried on the business of mining coal and making iron. Father handed down to son often-successful concerns. The Bagnall's, Gibbons', Hickman's, Jones' and Williams' were families who all made, and sometimes lost, fortunes in the business.

Furnaces erected on the coalfield initially used local supplies until they were exhausted. At first there was little, if any, canal traffic for the stone, but this slowly increased from 1790. Zachariah Parkes and John Wilkinson were both early carriers of ironstone by canal to their works. The volume of traffic then increased in relation to the number of furnaces at work and there was a significant rise in the amount of ironstone carried between 1800 and 1830. This fact is demonstrated by the several references in Birmingham Canal minute books to ironmasters making applications for the alteration or reduction of carriage rates.

It is difficult in many cases to differentiate between coal and ironstone transport when looking at specific mines, since collieries could raise both coal and ironstone. Some mines were, however, noted for their ironstone production and ironstone trade can be more accurately predicted. James Bridge, near Walsall, was a particularly productive mine. Ironstone raised in pits on this colliery was taken by tramroad to a basin beside the Walsall Canal. Thomas Price and later John Bagnall & Sons took stone from here to be smelted in their blast furnaces.

After 1830 requests for reduction of BCN tolls became much more numerous. In April 1832 John Dawes attended the proprietors meeting to discuss rates of tonnage for coal, coke and ironstone to his Oldbury Furnaces. There were then differential rates, on the BCN, for coal transported over three miles and ironstone over four miles. In July 1833 contracts were arranged with George Parker (Coneygre Furnaces), H.B. Whitehouse (Wallbrook Furnaces), Philip Williams (Wednesbury Oak and Union Furnaces) for ironstone travelling less than four miles on the canal to pay a base rate. Several furnaces owners on Dudley and Stourbridge Canals also requested, during 1833, a reduction of tonnage of ironstone passing from the Birmingham Canal into the Dudley Canal. By September 1833 contracts had been arranged with the British Iron Co. (Dudley Wood and Netherton Furnaces), Mr Firmstone (Lays Furnace) and Mr Gibbons (Corbyns Hall and Level Furnaces).

Not all furnaces were located beside the canal, there were several at a distance from the waterway. In these cases tramways or railroads provided the means of transport between the two. Tramways were thus directed from mines to furnace and furnace to canal basin. Tramroads criss-crossed the barren landscape to bring their cargo of minerals to the furnace bank. The mines near Wolverhampton were in a particularly alien environment. A forest of chimneys covered the skyline. Poles and pulleys were everywhere to direct the engine ropes to the pitheads.

The Priestfield Furnaces were built about 200yds from the canal and were connected by tramway to a wharf beside a canal basin. Tramways also were directed in a north-easterly direction to coal and ironstone mines on the Priestfield Colliery. The greatest expansion was during William Ward's time when the full extent of the mineral property was developed. As the mineral supply was exhausted near to the furnaces, the tramway was extended to new property. The length of lines gradually increased until the outskirts of Willenhall were reached. By 1846, in addition to Priestfield, William Ward was working mines on the Meadows and Portobello Collieries.

Coal and ironstone extraction proved a profitable venture on the land known as the 'Parkfields Estate' that comprised about 100 acres. Buried below the surface were seams of Roughhill Ironstone, Blue Flats Ironstone and seams of coal. The prolific ironmaster Samuel Fereday, in partnership with Richard Smith and others, came to develop the estate during the early 1800s. Fereday's bankruptcy that came with the iron trade depression of 1816 and 1817 led to the sale of the Parkfields Estate which was advertised in 1819, 1820 and later to find a suitable buyer.

The Parkfield Ironworks were first established in 1826. They were named after the nearby Park Hall Collegiate School, an old and established Catholic foundation. It was the practice hereabouts for the land around a particular feature to adopt the suffix 'fields'. So there was Parkfield, Millfield (named after the nearby flour mill), Priestfield and Horseleyfield.

Coal and ironstone mining on, or near, the Parkfield Estate was well established before 1810. Fereday & Co. were then working mines south of the turnpike road. Gibbons and Bickley had the Cockshutts Colliery, north of the turnpike, while W.H. Smith worked pits near Catchems Corner. All these mines would later become part of the Parkfield Estate.

There were always those willing to try their hand in this risky business. John Underhill, George Bishton and William Picken took a lease of the 45-acre Cockshutts Colliery in 1821. They were aided in this venture through George Bishton's appointment as agent for the Shropshire estates belonging to George William, Lord Stafford in 1822. Further mineral properties were subsequently acquired. A small 4-acre mineral estate, called 'Hensbirch', was leased in 1824 to George and Thomas Bishton and John Underhill. This acquisition was followed in July 1825 by the lease of the 100-acre Parkfield Estate in the parish of Sedgley from Lord Stafford. It was on this Parkfield Estate that the Parkfield Ironworks was erected for Underhill & Bishton. The first two blast furnaces were put up during 1826, a third was added in 1827 and, finally, a fourth in 1828. There was a good supply of local ironstone and coal (raised from the Thin Coal measures). These minerals were raised from pits on the adjacent collieries, which extended from the works to the Birmingham Canal Navigations.

The term 'Thin' is used to describe the coal measures in these parts. The Black Country is better known for its Thick Coal mines, but in this part of Ettingshall the coal was found at New Mine (7ft thick), Fire Clay (7ft 6in), Bottom (11ft) and Mealey Gray. Thick Coal was only found on the far eastern edge of the Park Field Estate. Techniques for working these measures were different to those used for Thick Coal, the miners were also paid less for getting Thin Coal. The ironstone measures were found as getting Rock, Poor Robbins, White Ironstone, Ball, Blue Flats and Diamonds. The last two seams were good sources of iron and were frequently sought after by the ironmakers.

New Furnace Construction

From 1829 the works came under the control of the Parkfield Iron Co., which was a partnership of John Underhill, John Bishton, William Bishton and Thomas Bishton. George Bishton also retained a share in the estate. In order to finance the venture a mortgage amounting to £17,000 was secured from Ralph Adderley.

Lord Stafford's involvement was eventually to result in civil action by him against George Bishton during 1833. It seems George was negligent in paying certain monies over to Lord Stafford and the action was to reclaim the debt. George Bishton was also then involved in a partnership at the Langley Field Iron and Coal Works, Dawley, Shropshire. George Bishton had persuaded Lord Stafford to take share in both Langley Field and Parkfield. The resulting action led John Bishton to agree to pay the value of the share in Langley Field and Parkfield back to Lord Stafford.

George Bishton, from 1833, was involved in the management of the Capponfield Ironworks, near Bilston. Bishton had replaced Thomas Otway in the partnership with Harry Parkes there. The Capponfield Ironworks was located beside the old Birmingham Canal, about two miles from the Parkfield canal basin at Spring Vale. For the Parkfield Co., Capponfield provided a useful outlet for their pig iron. For here, at Capponfield, pig iron was converted in wrought iron and rolled into bars, rails, etc. The Parkfield Co. even leased the Capponfield Works for a year between 1834 and 1835.

Various proprietors had worked the Capponfield Ironworks prior to Bishton family involvement. Several, such as John Read and William Aston, had suffered bankruptcy proceedings while it was in their ownership. For the Bishton's the trend was no different. The *London Gazette* announced the bankruptcy of John Bishton, Dawley, in June 1835. William Chapman was authorised to receive all debts due to the estate of John Bishton & Co., Capponfield Ironworks.

James Foster, ironmaster of Stourbridge, took over the lease of the Capponfield Ironworks from February 1836. He also acquired the lease of New Bradley, near Bilston, following the bankruptcy of James Turley. In 1853 both works passed to the Chillington Iron Co., a company in which Foster was a major shareholder.

The court action brought by Lord Stafford against George Bishton had caused a heavy cost to John Bishton & Co. who faced bankruptcy proceedings, probably because of it. Meanwhile Parkfield Furnaces continued to be carried on by the Parkfield Iron Co. for which the burden of management fell to John Underhill. It remained in his control until his death and was then carried on by his assignees. A group of businessmen were charged with the control of the business, for which John Pugh was managing partner.

Each mineral estate had a limited amount of minerals to give. Parkfield was in an iron-rich district, but all the available minerals properties had been acquired by various ironmasters. To the south of Parkfield was the Ettingshall Park Estate that sent the ironstone to the Spring Vale Furnaces. To the north was the Wolverhampton and Rough Hills Estate, which supplied the Wolverhampton Furnaces, while Millfield and Priestfield Furnaces took stone from the estates to the east and north east of Parkfields. The extent of the mines for this period can be seen on the ordnance survey first published about 1834. The Parkfield Furnaces are shown at the centre of a tramway system that extended across the Parkfield Colliery from the *Fighting Cocks* to the Lanesfield and Ettingshall Lodge Collieries that lay between Ettingshall Lane and the canal. A branch of the tramway crossed the turnpike to pits on the Cockshutts Colliery while another branch of the

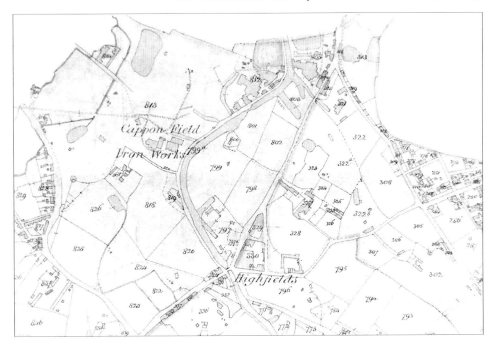

The Capponfield and Highfields area of Sedgley was extensively mined. The Birmingham Canal Old Main Line in this view passes from north to south. Various Ironworks are identified, including Bishtons Ironworks (812), Bagnall & Sons Ironworks and Furnaces (813) and Firmstones Highfield Ironworks (330). (Source: Sedgley Parish Plan, William Fowler & Son (1842), Edward, Son & Bigwood and Mathews MS, Birmingham Library Archives)

tramway ran south east to the canal basin beside the Spring Vale Ironworks. A similar pattern can be discerned from the Sedgley Tithe Map, published during the early 1840s. The ironmaking plant at Priestfield comprised three furnaces. In 1812 these furnaces produced 4,998 tons of pig iron. Various improvements were made to the plant, including newer and better furnaces. Nos 2 and 3 were replaced in 1829, while No.1 was rebuilt in 1836. By 1839 the annual make of iron had increased to 10,264 tons.

The Fereday links with the local iron trade were further eroded when John Turton Fereday severed his connections with the industry. During August 1832 the Bradley Ironworks was advertised to be let by private contract in possession of Joseph Turton Fereday, consisting of two blast furnaces, three refineries, a forge and mills capable of making 250 tons of bar, rods, hoops, sheets and boilerplate. They then passed to a partnership of ironmasters trading as John Wilkinson & Co.

Joseph Fereday was in partnership with John Jones, first at furnaces at the Buffery and then Windmill End. The Bancks family, as noted in Chapter Two, had been associated with the single furnace at Windmill End since its construction, although partnerships for working this furnace changed. Lewis's *Directory* for 1820 lists Hughes, Horton and Downing as proprietors, but by 1823 control had reverted back to William Bancks and it is he that is mentioned in sales advertisements for the furnace and foundry. They were then capable of making 50-60 tons of pig iron per week. The plant also included air furnaces, a finery, a moulding shop and a three-storey warehouse. The chief products were castings.

Two new furnaces were made at Windmill End during 1825 according to Scrivener in his account of the iron trade. This development was apparently due to Fereday & Jones. By 1828 they were under distress for rents and royalties owed to the landowners and a sale of colliery whimseys, gins, pit ropes, pig iron and casting was ordered to recoup the debts. John Jones was also later in partnership with John Blackwell at Buffery, and Blackwell and Jones together with Thomas Oakes established the Ketley Furnaces near Shut End.

Some furnaces remained small undertakings. Blast furnaces were established on the outskirts of Willenhall, beside the Wyrley & Essington Canal, in around 1800. Coal and ironstone were worked in gin and whimsey pits on either side of the waterway on an estate that covered 60 acres. There was a canal basin and a small foundry on the site. A pumping or water engine was also erected to drain the mines and a brick kiln provided material to line the mineshafts.

By April 1808 there were two two-blast furnaces near the village of Pool Hayes, complete with a blast engine and five workmen's houses, all then in the ownership of Mr Penn. By 1825 these furnaces were known as 'Coltham' and were owned by a Mr Tristham. In addition to the furnaces the mineral estate comprised 60 acres with measures of coal and ironstone, that is: Top Coal, Bind Stone, Flying Reed, Gubbin Stone, Heathen Coal, White Stone, Sulphur Coal, Yard Coal and Bottom Coal, Ball Ironstone, Gubbin Ironstone and Blue Flat Ironstone. There was then a period when the Coltham Coal & Iron Co. had the estate, with Daniel Bagnall as agent.

Richard Mainwaring was the proprietor of Coltham Furnaces in 1839 and also had a brief interest in part of the Highfields Ironworks near Bilston. Richard and his family had a long history of mining; first in Wroxwardine, Shropshire, and later for the Earl of Dudley at Netherton, where they were employed as ground bailiffs. It did not equip him for the management of a blast furnace, however. In 1840 Richard Mainwaring was made bankrupt and the works were sold. Coltham Furnace was dismantled after 1842. John Bagnall & Sons worked the coal and ironstone measures until 1852. At which time there were still measures of yard and Five Feet Coal, Gubbin, Brown and Blue Flat Ironstone unworked. Bloomer & Davis later mined these minerals. They benefited by the construction of a canal company tramway, the Short Heath Tramroad, which was constructed about 1860.

The Earl of Dudley owned the largest mineral estate within South Staffordshire and East Worcestershire. Generations of iron and coalmasters had exploited the resources, but there was still opportunity for those in the trade to get the minerals and work the remaining ironstone. All were in the hands of firms who paid rent to the estate and also a royalty on iron made.

The death of the Earl of Dudley had left the estate in the hands of trustees, including the powerful Earl of Stafford. Francis Downing was agent for the estate, which comprised the following properties.

Collieries and Ironstone Mines
Bilston
Brierley Hill
Brockmoor
Coneygree
Dudley Wood
Foxyards
Gornal Wood
Kingswinford
Level
Netherton
Parkhead
Princes End

Ironworks and Furnaces
Coneygree Furnace
Cradley Ironworks
Deepfields Furnaces
New Level Ironworks
Old Level Furnaces and Ironworks
Parkhead Furnaces

Others
Kingswinford Railway
Limestone Mines and Limeworks
Round Oak Timber Yard

George Parker then leased the Coneygree Furnace, while J. & B. Gibbons paid a rent for the New Level Furnaces as well as royalties on the iron made and the use of the blast engine. At the Level Ironworks the Gibbons continued to lease the forge, while William Izon rented the 'Old Level' Furnaces. Samuel Evers leased Cradley Forge and Ironworks and the partnership of Evers and

Martin had replaced Zephaniah Parkes at Parkhead Furnaces. Robert Martin paid the rent of the Parkhead Foundry and the royalty on castings made there. Cradley Forge had a long connection with the iron trade, being worked by Dud Dudley, the Foleys and the Knights. The original blast furnace, erected during the seventeenth century, was said to have lasted there until the 1830s.

A report on the estate was commissioned in 1835, which was to have an important influence on the future of the way ironmaking and coal mining was to be conducted. The report was compiled by Smith and Liddell and examined all the works within the estate. They made a general survey and included their observations in a book of inspection. Their general observations included geological data and show the extent of their knowledge about the coal and ironstone seems at that time, as well as pointing the way for further exploration.

It was accepted that the district contained beds of coal ironstone and limestone of various thickness and quality. Many of the beds which were peculiarly adapted for making and manufacturing iron have been opened and extensively worked over a considerable portion of what has heretofore been generally considered the limitation of the coal district. The Thick, or Ten-Yard, Coal having been always preferred for domestic and manufacturing purposes had been extensively worked and wrought so that little remained in the parishes of Bilston, Wednesbury, Sedgley, Tipton and Dudley compared with the quantity which those parishes once contained.

The increased demand for iron and manufacturers generally, the growing use of steam machinery and the extension of inland navigation within the previous forty years had produced an extraordinary call for the raw material and given a powerful impulse to mining speculation and industry throughout the district. Unfortunately, it was at the expense of the rapid exhaustion of Thick Coal measures. This in turn led to the search for lower mines, which were eventually discovered, consisting of a variety of thinner coal beds mixed with valuable seams of ironstone and fireclay. Working the Thick Coal, and other measures, had gradually provided better knowledge as to the extent of the available seams, that is, the measures which could be reached by the equipment of that time. Studies in geology had led to a greater understanding of how the strata was formed and predictions could be made as to where minerals might be found. In this way, mining engineers were able to compare the order of coal beds on either side of the limestone hills around Dudley and Sedgley and deduce where unexploited minerals might be.

Smith and Liddell went on to make detailed examination of all the mineral property and identified the areas where mining was still productive and made recommendations for development. This report was followed by Richard Smith's appointment as agent, in 1836, to replace Francis Downing.

When Richard Smith returned from Nova Scotia, his mining skills were again in demand. He had a champion in of the Earl of Stafford, who was an executor of the Earl of Dudley's estate. Richard set about a programme of modernisation, taking back certain undertakings to be run directly for the estate. The first project was at Coneygree, that was still leased by George Parker. This furnace had consumed all the minerals produced in the mines owned by Parker. Work had started to make a second blast furnace, which when erected was calculated to consume 24,250 tons of coal 15,000 tons of ironstone and 4,000 tons of limestone every year. Smith and Liddell were informed, in 1835, that Mr Parker intended to quit this furnace and that Payton and Hopkins had considered becoming tenants of the ironworks, drawing on the minerals in their part of Coneygree Colliery. Surviving records indicate that Payton and Hopkins were working the Coneygree Furnace in 1836. Coneygree did not remain in their ownership for long, however. Smith arranged for their management to be directly controlled by the estate and not an independent ironmaster. New Level Furnaces were also taken back into the fold and were completely reconstructed under Smith's directions. Deepfield, Old Level and Parkhead were retained by the leaseholders.

The split ownership of works at Old Level Furnaces at Brierley Hill continued for a number of years. In September 1837, William Izon signed a new lease that describes in detail the Old Level Furnace property. The two blast furnaces were constructed of red and white bricks and held together by cast and wrought-iron binders. A beam engine, which had a 36in cylinder and a 24ft-long beam, was employed to create the blast. There was also a foundry and casting-house, each with a tiled roof. A bridge was constructed across the canal to serve an adit. The tunnel passed under the centre of Brierley Hill where the Izon's mined the coal and ironstone measures.

Furnace Design and Improvement

The 1830s proved to be a time of consolidation, but also of improved furnace design. John Gibbons, a member of the family that managed New Level and Corbyns Hall Furnaces made some fundamental observations concerning furnace design. In the early days of the nineteenth century the furnace hearth was still mainly square. It was a legacy from earlier times when it was made of stone blocks. Gibbons speculated that when a relined furnace was blown in it would be several months before it reached normal production. He observed blown out furnaces and found that the square corners of hearth had worn off, leaving it roughly round.

In 1832 Gibbons decided to see what would happen if he built a new type of furnace. His furnace, with a round hearth, was built alongside a traditional one and a trial of both furnaces was made over a six-month period. The old furnace put out about 75 tons of iron per week, which was about average for the period. The new furnace achieved the unprecedented figure of 100 tons. John Gibbons made further experiments with furnace design through altering the angle of the bosh, increasing the hearth size and the height of stack, all with successful results. Gibbons also introduced a new material to furnace charge. All over the area were groups of waste heaps from puddling furnaces and refineries, these heaps contained an iron-rich slag locally known as 'cinder'. By taking this slag, breaking it up and adding it to the furnace charge, Gibbons found that the iron in the slag could be reclaimed. The percentage of slag to ore would vary from a half to two-thirds cinder, but the resulting iron made was still of good quality.

South Staffordshire blast furnaces initially employed a blast at the temperature drawn from the outside. In 1824 James Beaumont Neilson, manager for Glasgow Gasworks, patented a technique where the blast air was heated prior to entering the furnace. The heated air, when raised to a high temperature, improved iron production and the technique was adopted by several ironmasters. Trials were first conducted at the Clyde Ironworks where the air passing through heated iron pipes passed into the blast furnace. South Staffordshire ironmasters were slow to adopt the invention and perhaps this was a factor that inhibited their production compared to other furnace owners in the North and in South Wales. Lloyds Foster was the first to take up the hot blast method for their furnaces at Old Park, Wednesbury. They had a device for this purpose installed by 1835.

The concept of using a lining in the puddling furnace that would combine with the carbon and speed up the puddling process was developed during this period. Puddling was initially a lengthy process involving the furnace and the finery, but several ironmaster contributed to improvements of the process. In 1812 Jeremiah Dimmack patented some basic improvements to furnace design (patent 3,569), which regulated grate size and suggested the use of dampers in the chimney. Another development was the substitution of a sand floor for an iron floor in the puddling furnace. Daniel and George Horton patented an improved puddling furnace in 1832 (patent 6,299) that included slag and cinder placed along the bottom and sides of the furnace onto which the broken pig iron were placed. The temperature of the furnace was raised to a higher degree than normal and the iron was allowed to 'boil', while being stirred by the puddler.

Daniel and George Horton had found part of the answer but not the best material for a lining. Attempts to find a suitable lining were solved by Joseph Hall (1789–1862) of Tipton. He began making experiments on iron puddling as early as 1811. A partnership was formed between Hall and Richard Bradley and in 1830 they erected the Bloomfield Ironworks, where Joseph Hall continued his experiments. Federick Isaac Welch also joined this partnership, bringing additional capital into the venture, but he left again in 1834 and William Barrows took his place.

Bloomfield Ironworks, where Joseph Hall perfected the improvements to the iron-puddling process which contributed to the success of South Staffordshire ironworks. These works were located beside the Birmingham Canal Navigation's Old Main Line at Tipton.

In 1839 Hall took out a patent (7,778) for preparing 'tap cinder' for use in the puddling furnace. Hall's method involved the heating (or calcining) the cinder in a kiln. The mass produced was known as 'bull dog', which was broken down into small pieces and used as the furnace lining. Hall discovered that the iron oxide combined with the carbon and enabled the iron to be melted in one stage – this dispensed, finally, with the finery. The method was different to all previous modes of puddling: Hall lined his puddling furnace with iron oxide (from the bull dog) and boiled (like Horton), instead of baked, his pig in 4-5cwt lots at a high temperature in a bath of fluid basic oxide. In all previous processes the fluid oxide was necessary for removing carbon from the pig and had been obtained by oxidising the iron itself; but by supplying iron from outside, Hall avoided this source of loss. Also, while Cort's process removed silicon and carbon, Hall's also reduced the phosphorus and sulphur, giving a much superior material. Hall's process established the reputation of South Staffordshire Bar. It was a significant discovery that made an important contribution to the iron trade.

Joseph Hall provided an explanation of how his discovery was made in an account published in the *Birmingham Journal*, 27 September 1857, where he attributed his discovery to a long-standing interest in recycling iron and iron scrap. Following the example of his employer, Hall would gather scrap iron to add to the refinery. From 1816 this included the contents of the puddler's boshes. This was sieved to collect pieces of iron, left over from the puddling process. Hall recounted that some of the puddling furnaces were changed between the years 1819 and 1820 to include large iron bottoms laid with cinder instead of sand and this design was adopted at his place of work. Unfortunately, this method created much more waste which was impossible to separate by the sieve. Joseph Hall decided on another course of action. He saved up the bosh slag until the

The Newcomen Engine provided the first form of steam power to drain mines. This type of engine is referred to as an atmospheric engine as it relied on air pressure to bring down the beam after an influx of steam had raised it. In this engraving, a Newcomen Engine is shown to be adapted to turn a flywheel as may suit an ironworks. Local engineers were very inventive and had the skills to adapt and repair old engines as well as produce new engines.

Saturday night, when the puddlers had finished their week's work. He then charged a furnace with the slag and increased the heat in order to fuse it. In what was probably the first attempt at boiling iron, Hall became alarmed at the way the furnace filled up and overflowed. His initial thought was that the experiment was a failure, but afterwards Joseph retrieved from the cinder pieces of iron that could balled up for the hammer. This iron proved to be some of the best he had worked with. The discovery that this iron had come from furnace waste deserved further investigation.

Hall's experiments of Saturday night led him to spend years of research that culminated with the patent of 1839. He looked at the three methods, which were the old dry puddling method, fusing bosh slag and the pig boiling system, and became convinced that iron made from dry puddling was partially spoiled. He was particularly concerned about the large amounts of cold water used by the puddler, which was added for the effect of drying the metal quickly, hence the term 'dry'. Having discovered a new method of puddling, the next step was to redesign the furnace, replacing the old type of brick and clay with a new type enclosed with plates. But there was still the need of a furnace lining for protection. Hall fixed upon the bottom cinder produced by refining and from this substance known as 'tap cinder' came the material mentioned in the patent.

Ironmasters used the term 'wet puddling', or 'pig boiling', to distinguish between Hall's method and the former methods, which were frequently termed 'dry puddling'. The older two-stage method of 'dry puddling' was effectively obsolete in the Black Country by 1850. The saving of time was of particular importance in gaining new orders such as the supply of rails and other ironwork to the growing national railway network.

CHAPTER 5

FURNACES AND IRONWORKS
TRADE EXPANSION, 1838–1855

Again, iron came into increased demand with additional requirements for cast and wrought iron. Ironmasters tended to specialise in one or the other. Firms like Henry Bickerton Whitehouse and William Ward produced iron for casting purposes and also made castings themselves. Other ironmasters continued with the supply of pig iron for working up into wrought iron. These included firms like John Bagnall and Philip Williams & Sons. The new method of puddling provided an important step forward for the later group at a time when an increasing demand was being made for new railway schemes. In addition to pig iron manufacturers there was also an increasing number of ironmasters who were content to purchase pig iron and work it up for specific uses. Some made their money from taking any orders to produce what was called 'merchant iron', others made a specific range of products; bar, rods and sheets. Some produced hoop iron to cater for the cooperage trade and some made boilerplates for firms who made boilers. There was still a considerable demand for nail rods, but the days of the family nailers and the nail ironmongers that supplied them were numbered. Increased mechanisation in the industry and the introduction of cutting nails from sheets meant that new markets were opening up at the expense of the old.

The new ironworks tended to be built along the canalside, taking up the free spaces left through early mining or the fields and farms hitherto left untouched. Gradually the canal became lined with works and the waterway became busier with boats carrying coal, pig iron and finished products. The water-powered mills were in decline. Forges and ironworks that lacked a convenient canal link were either closing or taking on other uses such as the production of edge tools. Exceptions to this trend included the Stour Valley Ironworks along the Staffordshire & Worcestershire Canal. Lee & Bolton, during their ownership, made considerable improvements and enlargements to the Hyde Works, near Kinver, and Williams & Warden made similar improvements to the Whittington Works.

Local ironstone continued to supply ironmasters, in a pattern of events that was typical of the period. As ironmasters exhausted local supplies, others were acquired from mines further away and the canal was usually their means of transport. Gradually the whole coalfield became a mosaic of pits in work, separated by patches of worked out and unworked minerals. As coalmasters and ironmasters alike searched for new mineral property, a proliferation of tramways were made to canalside wharves. Each patch of ground was plundered for coal and ironstone, even the canal was not safe; mine owners frequently sought to work the minerals underneath.

Mineral transport along the canals from pithead to furnace steadily increased as local mines became exhausted. This increase was most marked between 1840 and 1860 when the greatest number of furnaces was in blast. Some firms relocated their furnaces to obtain fresh supplies of minerals. Apart from the local ironstone, ore and ironstone was brought from Lancashire, Northamptonshire, North Staffordshire, Oxfordshire, Shropshire and Warwickshire.

Canal Trade and Longweight System

Shropshire ironstone traffic was the first ironstone transport made over a distance. Ironstone mine owners sent stone out by trow, but the trade increased with the completion of horse towing paths on the River Severn between Stourport and Coalbrookdale. Narrow boats could then make the journey to Coalport and Broseley with relative ease. Such was the impact of this trade that the furnaces on the western bank, such as Barnetts Leasowe, Broseley and Calcutts were closed and the high-quality ironstone sent down the Severn to Stourport. From here, the narrow boats navigated the Staffordshire & Worcestershire Canal and Stourbridge Canals to reach the furnaces at Brierley Hill, Dudley and Netherton.

The Warwickshire ironstone was mined at Bedworth and Hawkesbury and taken along the Coventry Canal to Fazeley and then onto the Birmingham Canal network of waterways to reach various furnaces. Surviving Coventry Canal letterbooks reveal that the trade commenced during the 1840s and that there was significant traffic in ironstone along this waterway during the 1850s. Canal carriers travelling south to London and other locations along the line of the Grand Junction and Oxford Canals frequently arranged to collect ironstone on the return journey; this led the Oxford Canal Co. to make a charge for empty boats passing along their canal.

An early user of Bedworth stone was the firm of Colbourne & Groucutt of Broadwaters Furnaces near Wednesbury. In 1849 David Groucutt wrote to the Coventry Canal with the hope that they might charge tonnage longweight:

20 February 1849
Broadwaters Furnaces

We are serious to know if you can oblige us by charging your tonnage longweight as we are now doing a large scope of business on your canal and find the ironstone sellers go by your agent's gauge. We think it would be more business-like to charge the ironstone longweight as they do on the Birmingham Canal. We find your agents gauge the boats very near and whatever the charge on the ticket we have to pay not only for the tonnage but the ironstone also.

PRO Rail 818 Coventry Letterbooks

Mr Blackwell of Russell Hall Furnaces was another ironmaster who used Bedworth ironstone. In December 1850 he applied to the BCN proprietors for a reduction of the toll on ironstone passing from Fazeley to Russell Hall. The request was declined, but the traffic continued. Mr Blackwell must have considered this trade worthwhile since it involved navigating the Birmingham, Coventry, Dudley and Pensnett canals as well as the Dudley Tunnel to reach his tramway wharf. The stone then had to be transferred into tramway wagons and be carried another mile along the High Lanes Tramway to his furnaces.

Philip Williams & Son conveyed Bedworth ironstone to their Union Furnaces in West Bromwich. The tonnage charged by the Coventry Canal was disputed in January 1858. The manager of the Union Furnaces, P. Bennett, was concerned that shipments that he believed to be 20 tons were charged at 22 tons. The Coventry Canal proprietors replied that they were charged the tonnage per the weight their boats were gauged at. Their answer did not satisfy Mr Bennett who clearly expected that his loads would be charged longweight. In August 1858 Mr Bennett forwarded a payment for tonnage and complained that they would carry more traffic by the Coventry Canal were it not for the high tonnage rate and the extra weight charged on boats.

The whole issue of weight was a complicated one where interpretation of a ton was determined by the concept of longweight or shortweight. In the former, which was commonplace in the coal and ironstone trade, the weight loaded at the pit was often greater than the standard ton to partly compensate for loss en route. Canal companies such as the Birmingham, Oxford and Trent & Mersey took this practice into consideration and frequently exercised a degree of understanding

Lee & Bolton's Hyde Ironworks as seen from the Staffordshire & Worcestershire Canal, *c.* 1873. This engraving was published in Samuel Griffith's *Guide to the Iron Trade* (1873) and depicts an extensive ironworks dependent on coal supply from the canal. This structure replaced the earlier water-driven mills placed alongside the River Stour.

when gauging boats. The Coventry Canal Co. was less tolerant. Gauging was the principal means a canal company had of verifying the weight of cargoes in a boat and also enabled them to calculate their tolls for the traffic. Each commodity had a different rate and revenue might be lost if the canal company were to be generous with the charge. However, as Mr Bennett was quick to point out, ironmasters had the choice to determine by which means they carried and from where they selected their supplies. Canal Companies had to make the decision: which was the greater benefit, an accurate charge or increased revenue from more traffic.

Mr Bennett's response was based on the fact that he expected to pay a rate for 20 tons (that equated to 22 tons) and included the threat that he might source ironstone from elsewhere. There is an irony in the fact that the longweight practice lasted for so long due to the ironmasters, who saw true gauging as a surcharge on their transport costs and let it continue, particularly on the Birmingham Canal. Those purely involved in mining coal, the coalmasters, were definitely keen for more accurate measurement of boats. Despite memorials from coalmasters in 1831 and 1832 to improve gauging, despite the building of a station at Smethwick for fitting boat indexes, and despite parliamentary powers being granted to the Birmingham Canal Navigations to gauge boats in 1835, not all boats were gauged and the tolerance for longweight continued until a more sustained campaign led to the compulsory gauging of all boats from 1872. A contributing factor was the policies adopted by Frederick Smith, who succeeded his father, Richard, as agent to the Earl of Dudley. Frederick was keen to adopt the correct weight at the many mines then at work on the estate.

In the South Staffordshire and East Worcestershire district, the process of iron manufacture involved a complex mix of ingredients. Each furnace owner tipped a different mixture into his furnace. Local ironstones were frequently mixed with other ironstones and iron ores, particularly after the discovery of ironstone fields in Northamptonshire, Oxfordshire and Warwickshire. Boatloads of ore and stone would arrive from these districts to be crushed and mixed with local material and crushed cinder. The results were blends of pig iron with their own unique and special qualities. Ironmasters built up reputations with the different grades and qualities they produced, while those who purchased the iron adapted the different grades to their own special needs.

Further Improvements in Furnace Construction

Furnaces came be built using the improved design and yet more ironmasters became attracted to trying their hand in the district. Amongst the new entrepreneurs were Bramah & Cochrane who established Woodside Furnaces in around 1840. They became celebrated for their quality of pig iron, castings for boilerplate and tire iron.

The partnership comprised John Joseph Bramah and Alexander Brodie Cochrane. John Joseph Bramah, of Ashwood House, Kingswinsford, died in 1846. He was only in his forty-fifth year but had been fighting a long battle against a debilitating disease. In his relatively short life he had assisted the establishment of Woodside Furnaces and Foundry. He had also become a partner in the Bloomfield Ironworks (replacing Richard Bradley) and in the Horseley Co.

Alexander Brodie Cochrane was born in Dudley on 10 February 1813. He was engaged, at the age of seventeen, to work at Messr Grazebrooks collieries and ironworks near Dudley, then under the management of his father. In 1838 Cochrane started a foundry in Bilston with John Joseph Bramah and in 1840 commenced Woodside Ironworks with his father and Mr Bramah. When John Bramah died Charles Geach became a partner, as did Archibald Slate. Subsequently, Alexander was in partnership with his brother and son when the firm became known as 'Cochrane's'. The Cochrane family were also associated with colliery undertakings in Northumberland and Durham and the Ormesby Ironworks at Middlesbrough. Alexander Cochrane died on 23 June 1863.

Henry Bickerton Whitehouse built up an extensive business empire based on the coal and iron trades. Henry, the son of Benjamin Whitehouse and Sarah Bickerton (the daughter of

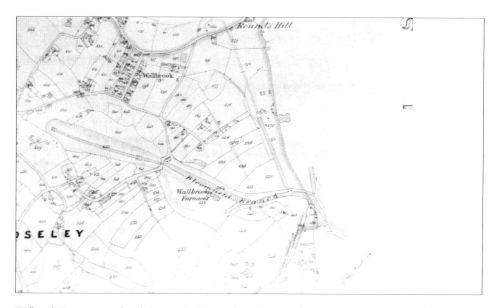

Walbrook Furnace was placed close to the Birmingham Canal ansd was served by a short branch known as the Deepfield Cut. With the widening of the canal there and the construction of Coseley Tunnel (1837-38), the foundations of the furnace were somewhat damaged. Iron production was transferred to nearby Priorsfield Furnaces. (Source: Sedgley Parish Plan, William Fowler & Son (1842), Edward, Son & Bigwood and Mathews MS, Birmingham Library Archives)

An engraving which shows Barrows & Hall's Factory Ironworks, *c.*1872. These premises were originally known as the Tipton Forge.

a Monmouthshire ironmaster), was baptised in Brierley Hill in 1796. He came to operate the Wallbrook Furnaces and collieries that were placed beside the Bloomfield arm of the Birmingham Canal. This length of waterway had been part of a scheme to tunnel under the ridge at Coseley to Deepfields, but the contractors had abandoned the project, leaving two sections at Bloomfield and Deepfield. It was not until 1838 that the tunnel was completed under a revised scheme engineered by Thomas Telford.

Whitehouse was in partnership with James Hunt until March 1830, when Henry took sole charge of the two furnaces at Wallbrook. The final construction of Coseley Tunnel between the years 1837 and 1838, as well as the widening of the canal, was an incentive for Henry Whitehouse to construct new furnaces alongside the Deepfields arm, where there were already located a number of basins and linking tramways to mines at Ettingshall and Sodom. Whitehouse was to erect three furnaces on an estate that became known as 'Priorsfield'. Wallbrook Furnaces were affected by subsidence following the canal-widening program; by 1842 they were standing idle and were later pulled down.

Henry Whitehouse had married Mary Smith at Tipton in September 1826. Their children included sons Arthur George, Benjamin, Edwin Smith and Henry Bickerton (junior). Benjamin was christened at Coseley on 2 January 1833. Parish records at this time describe Henry Bickerton Whitehouse as a coal and ironmaster of Ettingshall. Benjamin (1832–1913) was to help his father with the management of Priorsfield Furnaces, while Arthur and Edwin came to be associated with the H.B. Whitehouse & Sons venture at West Hallam Ironworks and Collieries that were served by the Nutbrook Canal. From January 1869 Arthur and Edwin Whitehouse traded as the West Hallam Coal & Iron Co. The younger Arthur came to live at High Lanes Farm, West Hallam, Derbyshire, where he carried business on as both an ironmaster and farmer.

Richard Bradley was agent to Horace St Paul and was responsible for his various holdings in Staffordshire that included Windmill End Furnaces, Willingsworth Furnaces, Pump House Colliery and Factory Ironworks. He managed and later leased the Factory Ironworks at Tipton, which was opposite Kier's Soap Factory. It was the soap factory that gave this area the name 'Factory'. The canal locks opened in 1838 as part of the improvements to the Birmingham Canal and became known as 'Factory Locks'. The bridge that carried the turnpike road (Hurst Lane) over the canal was known as 'Factory Bridge' and Bradleys Ironworks became known as 'Factory Ironworks'. Rate book records seem to confirm that Factory Ironworks was also known as 'Factory Forge' and 'Tipton Forge', formerly worked by Turton, Fereday and Firmstone. Bradley died in 1848 and the subsequent partnership of father and son, also Richard, was dissolved. Factory Ironworks was taken over by Barrows and Welch and worked in association with Barrow's other establishment, Bloomfield Ironworks.

At some locations, it was not unknown for different proprietors to work separate furnaces or parts of the ironworks plant. This became the case at Dudley Port Furnaces in 1839. John Turton Fereday established two furnaces near the turnpike at Dudley Port, but in February 1836 these had passed to Horton and Gill. This new business partnership comprised Daniel, George and Thomas Horton and Joseph Gill. The plant comprised two smelting furnaces, a casting house, a blast engine, a pattern shop and coke hearth. There was also part of a wharf beside the Old Birmingham Canal that could accommodate three boats. Coal and ironstone was obtained from Park Lane Colliery. In October 1839 Jesse Wright, Thomas Millership and Reuben Plant leased one of the furnaces and arranged for joint use of the blast engine. Millership and Plant then worked ironstone mines at Wednesfield Heath and wanted to produce pig iron from their stone, which was brought by canal to their furnace. This traffic lasted for almost a year. Their lease was surrendered in October 1840 when both furnaces passed into the sole ownership of Joseph Gill.

Daniel and George Horton were associated with various other enterprises, including boiler making. Daniel was in business at Hartshill as a boiler maker, when in about 1824 he set up as an ironworker with George Horton, who was currently residing in Liverpool. They established the Leys Ironworks and went onto lease the Russells Hall Furnaces and then Dudley Port. Financial problems subsequently led to them quitting these various ventures. Their demise came just as Bradley, Barrow and Hall were gaining recognition for the wet puddling process. Had the Hortons not been crippled by creditor actions, their contribution to the discovery of this invention might have been more widely recognised.

An event with a fortunate ending was the resurrection of the affairs at Corngreaves Ironworks. The legal action that crippled the British Iron Co. did not prevent the phoenix-like rise of the New British Iron Co. in 1844. The new management concentrated on iron-smelting operations at Corngreaves, eventually disposing of the older furnaces at Dudley Wood and Netherton. Two, four and eventually six furnaces were erected at Corngreaves. They also erected an extensive ironworks to compliment their other property at Nine Locks, Brierley Hill. Puddling furnaces and rolling mills were established close to the furnaces and served by their 3ft 2½in-gauge railway system that brought coal and ironstone from the mines and transported finished goods to the canal.

Like many places in the Black Country, Willenhall had mines of both coal and ironstone. The Thick Coal usually associated with mining in this region was absent in the Willenhall district, but there were seams of Bottom, Fire Clay and New Mine Coal, which were found here. The ironstone measures comprised Blue Flats and Gubbin Ball and both were eagerly sought by ironmasters to smelt into pig iron. However, with the exception of the area around New Invention and Short Heath, much of the mining property around Willenhall was slow to be exploited.

The 1841 Tithe Schedule identified few collieries within the parish. Mainwaring & Diggle were mining coal and ironstone around Coltham and their blast furnaces were close to the Wyrley & Essington Canal. Foster, Jones & Baker who traded as the Chillington Iron Co. had established mines at Moseley Hole, which were reached by the 3ft-gauge horse-drawn tramway from the Chillington Ironworks. There was also mineral property near the end of the Willenhall Branch Canal and three isolated mines known as 'Batting Hole', 'Bush Piece' and 'Little Monmore'.

Mining development was most evident by the canalside. Many ironworks and furnaces were located beside the waterway and canal boats were the principal mode of transport. The Walsall Canal passed to the south of Willenhall through Darlaston. Several important ironmasters had mineral property in Darlaston and neighbouring Willenhall. Ironstone was mined in a number of adjacent properties that lined the Walsall Canal from Wednesbury through to James Bridge.

The Willenhall Canal, which opened in 1803, was a short branch off the Walsall Canal near Herberts Park, which terminated in the south-west corner of Willenhall Parish. It first served mines owned chiefly by the Loxdale family but later served mines further away. Samuel Frost built a narrow-gauge tramroad for the Birmingham Canal Navigations that linked the Willenhall Canal to coal and ironstone mines at Bunkers Hill.

It was common practice at this time to sink shafts to get particular minerals. In any given area there might be a number of shafts sunk to different levels. The same shaft might draw minerals from a number of seams, but it was equally possible to have a shaft dedicated to draw from a particular seam of coal or ironstone. New Mine Coal could be extracted from one pit while Blue Flats Ironstone was taken from another. Moreover, each operation could be carried out by a different group of people. Butties, or chartermasters, often carried out the process of mining. It was they who employed miners to get the minerals for a royalty. The mine owners were the moneymen who usually provided the funds to sink the pit, operate the machinery and provided the pit bank staff. The situation could become very complex with different groups drawing minerals from the same set of mines.

The driving force that opened many mines in the Willenhall district was the search for ironstone; many of the collieries were established for the benefit of the ironmasters. Thus it was common to find the mines being worked for an ironmaking concern. Others appear to have been private ventures who would have sold the minerals for profit. George Bate worked pits on the Trentham Colliery up to 1865, when the estate was sold. These mines were sometimes called 'Little London' or otherwise, 'Noose Lane'. They were isolated pits that relied on wagons and carts to deliver their minerals. A common arrangement at these mines was a coal stage near a road where coal brought to bank was loaded into carts.

Thomas Davies & Son had the Bull Pleck and Neachills Colliery, which bordered on Bate's mines to the south. Bull Pleck lay near Noose Lane, while Neachills extended across Noose Lane into Wednesfield. These mines had land sale coal stages and surface railways, which were probably pit gauge. Thomas Davies had the Crookhay Furnaces at the end of the Dartmouth Branch and would have sent ironstone by canal to their furnaces.

Another mine owner at Little London was William Hanbury Sparrow, who traded under the title of Osier Bed Iron Co. The Osier Bed mines lay to the east of Bull Pleck and to the south east of Trentham and bordered on both properties. A narrow-gauge tramway took minerals from their mines down to the Osier Bed Furnaces near Moseley Hole. These furnaces lay in Bilston and were supplied with iron and coal from a number of mines. Samuel Frost operated another mine at Little London, which was also called 'Monmer Lane' by one Directory. The Chillington Iron Co. gradually extended its mineral property eastwards from Moseley Hole to Pool Meadow. They also continued in a north-easterly direction to Merrills Hole Estate in Wednesfield. These mines provided the essential ironstone for their Moseley Hole Furnaces, which had been erected in Willenhall parish during 1848.

Amongst the new furnaces established during this period were the Rough Hay Furnaces, belonging to Addenbrooke Smith & Pidcock and Herbert Park Furnaces belonging to David Jones. Rough Hay Furnaces were completed in 1843 along with a couple of pits. Their proximity to the new Tame Valley Canal, then nearing completion, was seen as an advantage for sending and receiving materials. The previous Addenbrooke partnership had been associated with the Moorcroft Furnaces, Moorcroft Collieries and associated Bradley Hall Colliery that were all linked by tramroads. During 1841 this estate was advertised for sale. The plant then included three blast furnaces, hot-air stoves, a forge and a mill. Much of this plant was either transferred or sold off and removed during 1842.

The new Rough Hay Furnaces partnership originally comprised Edward Addenbrooke of Kingswinford, Thomas Smith of Cheltenham and Charles Pidcock of Worcester. Articles were signed for this partnership on 27 August 1842. Management of the concern fell to Edward Addenbrooke and two of his sons; John (born 1817) and George (born 1825). Other sons included Edward, who took holy orders, and Henry and Thomas who were both solicitors. John Addenbrooke moved to Walsall where he made his home at the Elms, Sutton Road, and became principally involved with the development of the Rough Hay Furnaces.

The Herberts Park mineral estate comprised 18 acres of mines of New Mine Coal, fireclay coal, New Mine Ironstone and Roughills Ironstone. The mines were made available through the bankruptcy of William Marshall and Henry Rodgers in 1842. David Jones acquired the mines and erected Herberts Park Ironworks and Furnaces. By 1857 the ironworks possessed fifteen puddling furnaces. Jones also worked the Bridge Ironworks at Bradley, where he erected another blast furnace.

These were also changing times for the Horseley Engineering Works and Furnaces. During November 1845 the lease of these works was again for sale. The partnerships of York, Harrison & Co. and subsequently Bramah, Cochrane and Deeley had made improvements and additions. The plant now included foundries, air furnaces, cupolas and the two two-blast furnaces, erected between 1808 and 1809. There were boring mills, planing machines, slotting machines, screwing machines, lathes, pattern-making shops, furnaces and foundry yards, which had been used to make stationary and locomotive steam engines and structural ironworks such as canal bridges. The gasworks made gas to light premises.

Several changes in ownership occurred in the Bradley area; the birthplace of the South Staffordshire iron industry. The bankruptcy of George Jellicoe of the Bankfield Works, Bilston, and the Bordesley Ironworks in Birmingham in 1842 left Bankfield available for development by others, after a brief ownership by William Baldwin & Co. It came to be sold to Colbourne & Co. Between Bankfield and nearby Pothouse Bridge various ironworking establishments were erected. These included Bovereaux Iron and Tinplate (Baldwin & Co.), Bradleyfield Ironworks and Pothouse Bridge Ironworks (David Jones) which all capitalised on the local Bilston industry that converted tinplate goods into trays and waiters.

The Bradley Ironworks, formerly owned by John Turton Fereday, passed to George & Edward Thorneycroft, Wolverhampton ironmasters. The Thorneycrofts had the Shrubbery Ironworks on the canalside south of Walsall Street and also worked the Gotherseley Ironworks on the Smestow and the Swindon Ironworks on the River Stour and Staffordshire & Worcestershire Canal. They also once ran one of the water-powered mills owned by the Knight family. Twin brothers George and Edward gradually withdrew from the trade, leaving it to a partnership, George Benjamin Thorneycroft & Co. Edward died in 1847 and Swindon Ironworks was disposed off in the same year. George went on to become the first Mayor of Wolverhampton. He died in 1853.

The construction of the Ridgeacre Canal and associated branches improved access to the coal and ironstone mines on the eastern side of West Bromwich. The mineral-rich district around Balls Hill, developed by William James, saw fresh investment with the building of the Crookhay Furnaces which lay at the end of the Dartmouth Branch. Thomas Davies was responsible for the working of Crookhay mines and furnaces. This Davies came from the Dudley area, where members of his family were glassmakers and nail ironmongers. He worked mines at Kingswinford before working Crookhay and prospered from the trade. Thomas moved to Handsworth to live at the Austins, where he was a close neighbour of Samuel Dawes at the Leveretts.

Not all canal schemes were completed, or even begun. While the BCN managed to get its proposals of expansion accepted, involving projects such as the New Main Line, Tame Valley, Bentley and Rushall canals, other projects were less successful. The proposed London & Birmingham Canal that promised to unite the Stratford Canal with the Grand Junction Canal had tremendous support from the ironmasters, who saw potential improvement in canal carriage for their goods. Unfortunately, no work began on this project. In May 1836 the prospectus for the Stourbridge, Wolverhampton and Birmingham Junction Canal was published. William Fowler, of Birmingham, was the surveyor. The provisional committee included Edward Dixon of Ashwood House and James Foster, ironmaster of Stourton Castle. It promised to improve transport links to an area only previously served by tramroads to the Stourbridge Canal. The route passed through a working coalfield rich in coal, ironstone and clay, where ten blast furnaces were already at work. One flight of locks was contemplated, along with one tunnel, about a mile in length complete with a towing path. The total canal was five miles and six furlongs long. Construction commenced but only some two miles as far as the Oak Farm Ironworks was actually completed. In its final form this level waterway from the Stourbridge Canal, completed in 1840, became known as the 'Stourbridge Extension Canal'.

Construction of the Stourbridge Extension provided incentive for new furnace construction to supplement three previous iron-smelting ventures. The Gibbons, who had developed the coal and ironstone mines at Corbyns Hall, had constructed Corbyns Hall Furnaces and made a tramway to the Stourbridge Canal at Brockamoor. James Foster and Henry Bradley were responsible for the construction of the Shut End Furnaces, which were served by the standard-gauge Shut End, or Kingswinford, Railway from their ironworks and furnaces to the Staffordshire & Worcestershire Canal at Ashwood Basin. The third was the isolated Lower Gornal Furnace, operated by Edward Crockitt. The Ketley, New Corbyns Hall and Oak Farm Furnaces were erected between 1835 and 1840.

New Corbyns Hall was promoted by the Gibbons family who, when their new works were completed, arranged for the letting of their old furnaces to Matthews & Dudley. Ketley was established by the partnership of John Blackwell, John Jones and Thomas Oakes and Oak Farm was set up by a new company, the Oak Farm Co.

The most ambitious venture was the Oak Farm Furnaces and Ironworks, which was a comprehensive and integrated scheme. The Oak Farm Co. was created in 1835 and came to be operated by the partnership of Alexander MacNaughton, John Walker, James Boydell and Charles Roper. The estate belonged to Sir Stephen Glynne of Hawarden Castle and the development of the minerals was made at the suggestion of James Boydell, who was estate agent to Glynne. The mines were near to those exploited by Dud Dudley for his third furnace. The Oak Farm Co. established coal, ironstone and fireclay mines, erected two blast furnaces and an ironworks, a foundry, engineering works and edge-tool works. They had warehouses not only for their finished product but also for steel brought in for the edge-tool trade. They also operated Wall Heath New Forge.

The Midland Mining Report

The iron trade was (and still is) a cyclic industry which has had both good and bad trading periods. Production peaked then declined as demand fell off. Iron manufacturing firms had to adapt to changing needs. Those that did not adapt either closed down or often faced bankruptcy, which shut them down. At times it was not a safe occupation to be employed in. In 1843 there was further cause for concern. By this time the number of blast furnaces in the Black Country totalled 139, but only half were in blast and others were likely to be shut down. The ironmasters reduced the wages of their employees, with coal miners being the worst affected. At this time many of the ironmasters owned coalmines. The cost of getting coal therefore affected the final price of bar iron. If ironmasters could not sell their iron at a profit, then economies were made elsewhere. Whenever the market price of iron fell, less iron was manufactured. Both iron and coal workers then went on short time. If there was a further fall then the final economy was a reduction in wages. By the end of July 1843 the ironmasters had decided to cut wages. Thick Coal miners then received 3s 6d per day, while Thin Coal workers were paid 2s 3d per day. Thick Coal miners' wages were to be reduced by 6d and Thin Coal miners by 3d.

The inevitable strikes followed but the wage reduction was held. It was a particularly hard time for the colliers because many were on short time working as well. Two or three days' work was not an uncommon occurrence. By October 1843, however, the price of bar iron had gone up in the market place and the miners' wages were restored.

The Midland Mining Commission reports made during the years 1842 and 1843 provide much useful information about this period, which was labour intensive. Thomas Tancred singled out an example of a furnace producing 85 tons of iron per week. His table showed that 117 men and forty boys had a hand in the make:

	Men	Boys		Men	Boys
Furnace Men immediately connected with the Furnace			*Boat steerers, carters, etc. for coal, ironstone and limestone*		
Fillers	2	0	Steerers	1	1
Coke wheelers	2	0	Boat and cart unloaders	3	0
Limestone breakers	1	0	*Colliers, loaders, enginemen*		
Cokers	2	1	Colliers raising coal	16	6
Bridge stockers	2	1	Boat or cart loaders	3	1
Mine burners	2	1	Machineman	1	0
Loading cinders	2	1	*Miners, loaders and enginemen*		
Stock taker	1	0	Colliers raising ironstone	50	16
Blacksmith	1	1	Boat or cart loaders	2	1
Carpenters (job shared with colliery duties)	1	0	Engineers	4	0
Mine weighers (job shared with colliery duties)	1	1	Limestone getters and loaders	7	2
Cleaning coke hearths	0	1	Overlooker or clerk	1	0
Watchman	1	0			
Engineers	1	0	**Total**	**117**	**40**
Wheeling slack	1	0			
Machinemen	1	0			
Moulders	2	2			
Refiners	3	3			

So great was the concentration of furnaces and ironworks in a relatively small area of Britain that the whole area became known as the 'Black Country'. Smoky chimneys and fiery furnaces all helped to create the image. The canals served the corporate body that was the iron trade and was, at first, its lifeblood, until the railways came. Ribbons of iron came to challenge the ribbons of water as the premier carrier. It was, in fact, the building of the South Staffordshire railway that promoted the name, 'Black Country'. There have been some suggestions that the term was already in some local use, but with the completion of this railway to Lichfield the term 'Black Country' received wider publicity following the meetings that commemorated the opening in 1849. John Philip Dyott, Mayor of Lichfield, made a specific mention of the 'Black Country' in his speech, and journalists of the period adopted the term and came to be mentioned in their writings. In May 1850, when the South Staffordshire Railway was completed from Walsall to Dudley, the *Birmingham Journal* reported the advent of the first railway into and through the 'heart of the Black Country'.

Galvanised Iron Industry

This was a period of adoption of new methods and techniques. The increased availability of metallic zinc, otherwise known as 'spelter', had made it economic to coat iron with zinc to prevent rust. Such were the quantities of zinc that had become available, a structural use for it became possible.

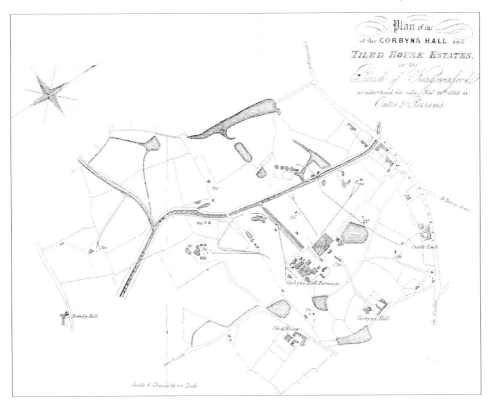

Above: Corbyns Hall Furnaces, 1849. Galvanised iron sales plan.

Right: Sales plan for Phoenix Ironworks, 1849.

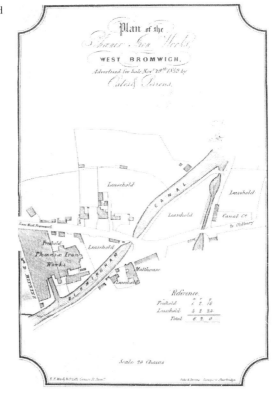

Zinc-coated iron became known as 'galvanised' and several local firms took the opportunity to exploit the new business.

Galvanised iron was patented and first marketed by the Galvanised Iron Co. They had two properties on the BCN: the Phoenix Ironworks at West Bromwich and collieries at Darlaston Green. BCN ledgers indicate that they paid tonnage during the 1840s. This firm also had the use of four blast furnaces at Corbyns Hall (belonging to William Matthews), near to the Stourbridge Extension Canal. Corbyns Hall was also the location of their galvanising works. The Galvanised Iron Co. ceased trading at the end of 1848 and their works were sold. The 1849 sales catalogue provides useful details about the firm's operation.

The company was clearly an important operation. The Corbyns Hall Estate possessed mines of coal and ironstone (Corbyns Hall and Tiled House Collieries) and the four blast furnaces were capable of making between 500 and 600 tons of iron per week. There were two foundries and a recently erected mill and forge. The Galvanising House dimensions were 105ft by 45ft. It contained picking tanks, a metal bath, two drying stoves and steam generator. The Darlaston Green property comprised six pit shafts that principally raised ironstone. Stone was conveyed in pit tubs along a tramway to an arm of the Walsall Canal. This short canal branch served other mines including those that belonged to Samuel Mills' Darlaston Green Furnaces.

The Phoenix Works were located beside the Balls Hill Branch near Swan Village and comprised mills and forges. Power was principally provided by two steam engines. A 100HP Boulton & Watt engine drove the 20in boilerplate train and the rail mill. A 60HP J. & G. Davis engine worked a forge, an 18in boilerplate mill, a sheet mill and a 16in train that made bar, T and angle iron. There were also thirty-four puddling and heating furnaces that were capable of manufacturing 350 to 400 tons of finished iron per week. Despite the break up of the Galvanised Iron Co., other firms continued with this trade, such as Morewoods and Tuppers.

Railway Mania

A public railway was first made through the iron district in 1837. This was the Grand Junction Railway, but the route avoided most of the major ironworks. A growing interest in railways led to proposals for many more. From 1844 details of new schemes were appearing almost weekly in the local press. Speculators rushed to invest in both the practical and impractical. For the ironmasters the opportunity of improved transport encouraged support. Certain ironmasters such as Thomas Badger and George Benjamin Thorneycroft lent their support to the Oxford, Worcester & Wolverhampton Railway that promised to serve an important slice of the iron district from Stourbridge, through Dudley and Tipton, to Wolverhampton. In April 1845 the Birmingham Canal Navigations Co. put forward a scheme that followed the banks of their waterway between Birmingham and Wolverhampton. Thomas Badger, William Hanbury Sparrow, G.B. Thorneycroft and Philip Williams all spoke in favour of the scheme which became known as the 'Birmingham, Wolverhampton & Stour Valley Railway'.

The Stour Valley Railway cut through the centre of the industrial heartland. Promotion of this railway was widely supported by local ironmasters and many signed a memorial for its construction. Notices of the provisional committee, in 1845, included the following coal and ironmasters:

James Boydell, Ironmaster, Kingswinsford	James & James Avery Griffin, Wythymore
Thomas Bolton, Ironmaster, Hyde Ironworks, Kinfare	Richard Heath, Ironmaster, Stourport
J. Joseph Bramah, Ironmaster, Woodside Ironworks,	Joseph Hall, Ironmaster, Tipton
J. Blackwell, J. Jones & T. Oakes, Ketley Iron Works	John Hartland, Ironmaster, Tipton

S.H. Blackwell, Ironmaster, Russell Hall Ironworks

Benjamin Best, Ironmaster, Wythymore near Dudley

Edward Bagnall, Coalmaster, Wednesbury

James Bagnall, Magistrate, West Bromwich

Richard Bagnall, Coalmaster, Tividale

William Bagnall, Ironmaster, Golds Hill

Enoch, Pearce & W.H. Baldwin, Ironfounders, Stourport

William Barrows, Ironmaster, Bloomfield Works, Tipton

Edwin Bullock, Iron manufacturer, West Bromwich

Joseph Bennitt, Coalmaster, Oldbury

William Bennitt, Magistrate, Dudley

Samuel Bayley, Coal & Ironmaster, Wednesbury

Edward Cresswell, Ironmaster, Edgbaston, Birmingham

Samuel Dawes, Ironmaster, West Bromwich

Charles, James & Samuel Evers, Ironmasters, Cradley

James Foster, Magistrate, Stourton Castle

William Orme Foster, Ironmaster, Stourbridge

Thomas Jesson, Coalmaster, West Bromwich

William Izon, Ironmaster, West Bromwich

Archibald, Samuel & Timothy Kenrick, Ironmasters

Edmund and T.W. Lewty, Wilden Ironworks, Stourport

Richard Mountford, Iron manufacturer, Coalbournbrook

J. & W. Millington, Ironmasters, Tipton

Joseph & Thomas Parkes, Ironmasters, Hyde Works

Edwin Pemberton, Ironmaster, Edgbaston, Birmingham

John Rogers. Ironfounder, Kidderminster

William Routh, Ironmaster, Corngreaves

W.H. Sparrow, Ironmaster, Wolverhampton

B. & J. Williams and W.M. Warden, Whittington

John Williams, Union Ironworks, West Bromwich

Philip Williams, Wednesbury Oak Ironworks

Richard Williams, Brierley Works, Kingswinford

Walter Williams, Ironmaster, West Bromwich

The Stour Valley Railway became part of the London & North Western Railway network and formed a direct link between Birmingham and Wolverhampton. The Birmingham, Wolverhampton and Dudley Railway was a rival route that received the support of the LNWR's competitor, the Great Western Railway. During 1845, this railway scheme was favoured by Michael Grazebrook, then chairman of the South Staffordshire Iron Trade. Other supporters included Thomas and William Bagnall, William Baldwin (Bilston Ironworks), John Barker (Chillington Ironworks), Thomas Davies (Crookhay Ironworks), E.B. Dimmack, James Foster, William Matthews (Corbyns Hall Ironworks), Sampson Lloyd (Wednesbury Old Park) and G.B. Thorneycroft (Shrubbery Ironworks). The Shropshire Union Railway & Canal Co. had a number of ironmasters as directors, such as John Barker, William Orme Foster, George Jones of Spring Vale Ironworks, William Hanbury Sparrow, J. & E. Walker of Gospel Oak Works and Philip Williams of Wednesbury Oak.

Not all railway schemes were adopted. Many were whittled away at the parliamentary approval stage, others failed through lack of finance. The Dudley, Madeley, Broseley & Ironbridge Railway was a railway intended to link the rich iron industry of Coalbrookdale with the Black Country. Those that invested in this venture included William Bennitt, Benjamin Best, Alexander Brodie Cochcrane, Thomas Davies (Crookhay Ironworks), James Evers, William Haden (Dixons Green), William Tomkinson Riley and Thomas Vernon.

The Rise of the South Staffordshire Malleable Iron Industry

As the number of blast furnaces increased, the South Staffordshire and East Worcestershire ironmasters came close to rivalling the production of South Wales and Monmouthshire. If furnace numbers were counted, the West Midlands would have been the masters, but in terms of actual production the local ironmasters were handicapped by the size and capacity of some of their older furnaces. While they may not have excelled in production of pig iron they certainly became the leaders in finished iron.

In 1850 there were sixteen Black Country blast furnace owners who also produced malleable iron. Their ironworks were frequently located on separate sites and the canal was integral to moving pig iron from the furnace yard to the ironworks. Only seven of the sixteen owners had ironworks establishments adjacent to the furnaces, therefore saving on transport costs. The reason for this state of affairs was essentially historical. The making of wrought and bar iron was conducted in establishments that often changed hands, through bankruptcy or owners leaving the trade. Ironmasters who chose both to smelt and puddle iron, or simply work pig iron up to malleable iron, took over existing works when they came up for sale. Some, such as the Stour Valley Forges, had been built as water-powered establishments but had been converted to use steam power. Industrialists in these times were keen to preserve working mills, especially where a viable transport route existed, be it canal or turnpike road. Whittington and Hyde Ironworks, both near Kinver, mentioned in the above list, were Stour Valley watermills that were served by the Staffordshire & Worcestershire Canal.

By 1850 the local ironworks were capable of making over 10,000 tons of iron per week. In reality the actual figure was about two thirds of this amount. The *Mining Journal* published the following list of employed and unemployed malleable ironworks for 1850:

Works	Capacity tons	Present Production tons	Owner or occupier
EMPLOYED			
Gold's Hill, Wednesbury & Toll End	700	400	John Bagnall & Sons
Shrubbery, Swan Garden & Bradley	600	500	J.B. Thorneycroft & Co.
Stourbridge, Brierley & Brockmoor	600	500	John Bradley & Co.
Bloomfield & Tipton Green	500	500	Barrows and Hall
Wednesbury Oak	400	300	P. Williams
Spring Vale	400	250	George Jones
Horseleyfields and Bilston Mill	350	200	W.H. Sparrow & Co.
Chillington & Leabrook	300	200	John Barker & Co.
Albion	300	200	Walter Williams
Bromford	300	200	William H. Dawes
Corngreaves	300	150	British Iron Co.
Crook Hay	300	200	Thomas Davis & Sons
Level	250	200	Hall, Holcroft & Co.
Bromley & Brettel Lane	150	120	J. & W. Wheeley
Pelsall & Golds Hill (New)	150	120	Davis & Bloomer
Roway	120	120	E. Page & Sons
Greets Green	120	120	J. Hartland & Co.
Smethwick	120	120	Spittle & Downing
Spon Lane	120	120	Silvester & Jackson
Darlaston Green	120	120	Bills & Co.

Bankfield	120	120	Coulborne & Co.
Ettingshall	120	120	Banks & Son
Dixons Green	120	120	William Haden
New Dudley Port	120	120	Plant Brothers
Tividale & Dudley Port	120	120	F. Giles & Co.
Whittington	100	100	J. Williams & Co.
Hyde	100	100	Lee & Bolton
Other sundry works	500	500	Various owners
UNEMPLOYED			
Oak Farm	400		W. Gladstone & Co.
Phoenix	300		Galvanised Iron Co.
High Fields	200		J.P. Firmstone
Brierley Hill	200		British Iron Co.
Caponfield	150		James Foster
Old Smethwick	150		John Williams & Co.
New Bradley	120		James Foster
Factory	120		Welch & Barrows
Old Leabrook	100		W. Bailey
New Leabrook	100		J. Hartland & Co.
Other sundry works	200		Various owners

The greatest manufacturer of malleable iron in 1850 was John Bagnall & Sons. The origins of their industrial empire have been discussed in a previous chapter. By this date their industrial empire was one of the largest in South Staffordshire. Bagnall worked collieries and ironstone mines to supply their furnaces at Capponfield and Goldsgreen. Pig iron produced there was distributed to their ironworks at Golds Hill, Toll End and Wednesbury. There was an integrated network of furnaces, ironworks and mines that were linked by the canal network. At the end of 1844 another ironworks was added. The forge belonging to Messr Russell was acquired by Bagnalls and renamed Imperial.

Iron and iron goods were sent to other parts of the country by way of the canal network. This business prospered after Thomas Bagnall went into partnership with James Fellows. The Bagnalls definitely benefited from the partnership. Fellows carried for them during their most productive years.

Control of the family firm was then in the hands of brothers John, James, Thomas and William. Samuel Bagnall had died in 1832. John died in 1849 and Thomas Bagnall left the partnership in 1857 and retired from the iron trade. A major readjustment to the partnership of John Bagnall & Sons was made in September 1857. Thomas Bagnall's sons, John Nock, Charles and Thomas (junior) who also had been in the partnership chose to leave and pursue other interests. The three sons of Thomas then became involved in coal and ironstone mining at the Ward Colliery, Bloxwich, while Charles and Thomas junior set up the Grosmont Furnaces near Whitby. A factor in the split may have been religion. The Bagnalls had been staunch Methodists, but Thomas and his brother, Henry, followed the doctrine and worshipped at the established Anglican church. James and William retained control of the family firm, with Richard Bagnall as partner in some of the coal mines.

Philip Williams & Sons had ironworks at Wednesbury Oak. Both pig and malleable iron were manufactured here, the colliery also provided a local supply of both coal and ironstone. These ironworks were adjacent to the furnaces. Philip Williams used surplus equipment from the Napoleonic War to start the ironworks business. During 1820 the forges and mills were constructed and Philip Williams began to manufacture the Mitre brand of iron at these works. Sons Walter, John, Philip, William and Henry later assisted their father with the management of the furnaces, mines and ironworks. They ably guided the firm through a period of prosperity.

In 1829 Philip Williams & Sons purchased the Union Estate at West Bromwich, where collieries had already been established by the Union Coal Co. Two blast furnaces were erected on the Union Colliery close to the Union Branch Canal. Associated with these furnaces was a foundry where castings of up to 10 tons were manufactured. A number of other mines were worked in the Wednesfield and Wolverhampton area which produced both coal and ironstone, but it was the ironstone that was of greater value.

By 1849 Philip Williams also had the Old Birchills Furnaces at Walsall. These furnaces had been established in 1806 worked under the partnership of Stubbs & James. Old Birchill furnaces were leased to F.C. Perry in 1856, who renamed them Roughwood. Mr Perry's career as an ironmaster was brief, however. In 1860 he was bankrupt and his assets were offered for sale. Roughwood Furnaces remained the property of the Williams brothers but were never put back into blast and remained derelict for a number of years. Walter Williams, the eldest son of Philip, was responsible for the Albion Ironworks that was erected during the mid-1830s beside the Walsall Canal at Albion. He also had an ironworks at Great Bridge.

George Jones had an important malleable ironworks at Spring Vale, Sedgley. This ironworks was placed close to the BCN Main Line but was served by the branch canal to Parkfield Furnaces. Iron made at the Bilston Furnaces, formerly worked by Sparrow & Co., was supplied to Spring Vale by boat or tramway. George had acquired the Bilston Furnaces in around 1840 and supplied them with ironstone and coal mined on the Ettingshall Lodge property. Jones had first commenced business in South Staffordshire as a coalmaster at Priest Leasowe Collieries, near Bilston, and also had the Phoenix Foundry, Snow Hill, Birmingham. From 1832 Jones was a director of the Bordesley Iron & Steel Co., also in Birmingham. In 1823 he leased iron and coalmines at Blaina, South Wales, and also established the Coseley Furnaces in Staffordshire. John Barker assisted George Jones in the management of his works, and had married his daughter, Theodosia. Barker, Jones and James Foster (Stourbridge ironmaster) had leased the Chillington mineral estate in 1822 and erected blast furnaces and an ironworks in 1828 that were laid out to the designs of John Urpeth Raistrick. John Barker was also in partnership with Goerge Jones at Blaina Ironworks and Furnaces in South Wales from 1827 until 1835.

The Chillington Furnaces adopted quite a revolutionary concept of transporting liquid pig iron to the puddling furnaces. In this manner they avoided the cost of burning fuel to re-melt the pig iron. They patented the invention (patent 6,300) in 1832. The names of George Jones, James Foster, John Barker and John Jones were recorded on the patent. George Jones also adopted this process at his other furnaces, such as in Spring Vale, Bilston.

Coseley Furnaces were disposed of during the 1840s when George Jones embarked on another business venture. In 1848 a new enterprise was begun at Birchills beside the Wyrley & Essington Canal. The New Birchills Furnaces and Ironworks were established near Green Lane and on the opposite side of the road to Thomas Highway's land. It was an extensive property where Jones was to erect five blast furnaces. Coal was supplied by canal from neighbouring mines at Forest and Harden.

George Jones made his home in Wolverhampton where he raised his son, John. By 1840 George had moved to Shakerley House, near Albrighton in Shropshire, where he owned a large estate. John Jones, his son, later lived at Ruckley Grange, also near Albrighton.

Bilston Furnaces and Ironworks was an integrated operation. A railway, the Spring Vale Railroad, served the furnaces and crossed the Birmingham Canal to Spring Vale Ironworks and Foundry. Spring Vale Ironworks comprised an Old and New Side Forge. The Old Side Forge had a hoop mill and sheet mill, while the New Side Forge had puddling furnaces, rail mills and a sheet mill. Collieries were to be found at Bilston Furnace, Ettingshall Park, Rookery and Spring Vale.

In around 1855 George Jones, failing in health, gave control of his industrial empire to his son, John, who managed it for a few years. George Jones died in March 1857. Unfortunately, the iron trade was experiencing a period of depression at that time. Many iron-business ventures failed, including the Jones's. John Jones was made bankrupt in 1864, and the estate was split up. At this time John Jones had focused his attention on the management of the Birchills Furnaces. Estimated liabilities amounted to £132,500, while the gross estimated value of assets was £200,000. These include the extensive landed property in Shropshire, 80,000 acres of land in America and a mortgage on George Hudson's properties at Albert Gate and Whitby Docks. Creditors appointed Mr George Smith of the Shifnal Bank and William Hatton of the Staffordshire Bank as trustees. The Civil War in America was said to have had an adverse effect on Mr Jones' affairs.

Early hopes were raised that a new public company would run the Jones' estate. Unfortunately, several years would pass before the legal affairs were resolved and the integrated site was disposed off in a piecemeal fashion. The split up segregated the collieries, ironworks and furnaces between different owners. One colliery, Birchills Field, was also partly lost through the construction of the Wolverhampton & Walsall Railway. Birchills Ironworks, in 1867, comprised twenty-eight puddling furnaces, three sheet mills and two hoop and merchant mills. This site was neatly bisected into two halves. The Birchills Hall Iron Co. acquired the northern section, while the forges to the south passed to Benjamin Bunch, who renamed them the Staffordshire Ironworks. The five blast furnaces at Birchills served by an incline were finally reduced to two, operated by the Castle Iron Co. Castle Furnaces were not pulled down until 1890. The remainder of the Birchills Estate was sold off piecemeal. Thomas Checkley took charge of Forest Colliery and worked it successfully until his death, while Harden Colliery passed to Elias Crapper. The furnaces were acquired by Alfred Hickman in around 1866 and renamed Spring Vale.

John Jones's home at Ruckley Grange and Estate was quite extensive, comprising the Grange and Ruckley Wood with 378 acres of land, the mansion, farmhouses, game keepers' and labourers' cottages, pleasure grounds and lawns. The Jones family also owned the Neachley Estate occupied by G. Holyoake that was 355 acres with mansion house and pleasure grounds. Both lots failed to make their reserve price when offered for sale in 1867.

The Chillington Iron Co. was once an important ironmaking concern. The Chillington Ironworks and Furnaces were located at Wolverhampton – both iron smelting and finished iron production was accomplished on one site. Control of the operation passed to John Barker. Tramways brought ironstone and coal from local mines to the furnaces and also connected the site with the canal basin at Monmore Green. The tramway system was quite extensive and was gradually extended towards Willenhall in order to reach unworked mineral property.

For about twenty years (1848–1868), the Chillington Co. also operated three furnaces at Moseley Hole. The tramway system connected Moseley Hole with Chillington Ironworks and also railway sidings on the Grand Junction Railway at Merrils Hole. The Chillington Co. was one of the earliest users of railway transport in the district. A third set of furnaces was worked at Bentley, near the terminus of the Ansons Branch Canal. Bentley also possessed coal and iron mines which remained an important source of minerals until the early 1880s. The Bentley Furnaces had been built for the Earl of Lichfield at about the time of the construction of the Ansons Branch in 1830.

The *Mining Journal* published a detailed description of the Chillington Furnaces in 1869. There were then four blast furnaces at Chillington, 45ft high, of which two were in work:

> The make at these two furnaces is from 120 to 140 tons per week each. The ores used are the Gubbin and balls and blue flats mixed with Frampton ore. Half coal and half coke is used as fuel, coal is coked in open heaps and hearths. From 20 to 30 tons of coal is placed around a chimney 7ft. high having several openings at the lower part. The larger pieces of coal are placed next to the chimney, and so as to form a channel to it. As the coking process goes on, the heap is covered with ashes. Three days are required to complete the process. The native ironstone is calcined in a large kiln, iron-plated outside, about 40ft high.

Chillington was one of a select group of furnaces that preferred to use lifts instead of inclines for raising materials to the top of the calcining kilns and furnaces. The material was conveyed on a carriage that followed guide rails to the top. Counter-balancing weights provided the means to draw it up to the top. After the cargo was unloaded the carriage was filled with water and was then allowed to descend again by gravity. Remarkably, and perhaps perilously, these carriages carried a brakesman who could stop it at any time! It was far more common in contemporary Black Country blast furnaces for an incline to be used for raising material to the top of the furnace. Waggons containing the charge were drawn up to the level of the furnace mouth where there was a platform known as the 'bridge'.

Iron from the blast furnace passed to the ironworks in an economical manner at Chillington Ironworks. The *Mining Journal* text provides a graphic illustration of how this was done:

> In front of the blast furnace, seven puddling furnaces are built. Into which the iron from the blast furnace is run direct, every two hours or so, as required; this economises the use of coal in puddling, but somewhat impedes the working of the blast furnaces; the remainder is run into the sand beds. There are three forge trains at Chillington; three mills for rolling sheet iron; two mills for nail rods, one of these also used for merchant iron; and two mills for rolling hoop iron. There are two rotary horizontal squeezers, one helve hammer, a common squeezers. The most of the boilers are heated by waste gases from the puddling and balling furnaces.

The canal basin connected with the BCN Main Line near the turnpike bridge. It was an important shipping point for pig and finished iron manufactured at the ironworks. Chillington Basin had tramway tracks on both sides of a covered loading shed. There was also a boatyard at the end of the basin where the company vessels were maintained.

The Bentley Estate, near Walsall, was held under lease from the Earl of Lichfield. Chillington acquired it in 1857 when the previous tenant, William Riley, was made bankrupt. The colliery estate extended from the Ansons Canal towards the Wyrley & Essington Canal and there were mineral wharves on both.

When the Chillington Co. took over Bentley, there were two complete and two incomplete furnaces, along with a foundry shop and casting house. Although iron continued to be smelted onsite, the colliery property proved to be of greater value to the company. Money was invested in a new plant for mineral extraction. The firm possessed another three ironworks, in addition to the original plant at Chillington. One was situated at Capponfield and another at Bradley. Both were placed beside the old Birmingham Canal and had been owned by James Foster until his death. The third was at Leabrook on the Walsall Canal.

John Barker senior profited both in money and status through his work in the iron trade. In 1850 he was High Sheriff for Staffordshire. His son, George, came to live at Albrighton House and was a Justice of the Peace and Deputy Lieutenant for Staffordshire. John Barker died at his home, Cleveland House, Wolverhampton, in November 1852. He was fifty-six. Control of the Chillington Iron Co. passed to sons George and Thomas.

The Bloomer Family – Nail Ironmongers to Ironmasters

Both Bagnall and Barker provide examples of successful family concerns. Another was the Bloomer family. Boaz Bloomer (1801–1874) and his family became successful ironmasters during the 1850s. Boaz had started in the iron trade as nail ironmonger and was initially in partnership with other members of his family. He was the son of Benjamin and Ann Bloomer and had gained sole control of the nail ironmonger business in December 1836 when Benjamin and Joshua Bloomer left the partnership.

Boaz had married Catherine Hornblower in St Thomas' church, Dudley, on 15 December 1825. They had ten children: Caleb, Boaz, Esther Ann, Giles, George Bailey, Silas, Cornelius, Prudence Bailey, Sarah and Benjamin Giles, although Silas died in infancy. Some were baptised at St Thomas' church; later, the baptisms were conducted at King Street Weslyan Methodist chapel in Dudley. Through the Wesleyan church Boaz made some useful connections, including the Bagnalls and Thomas Davis. Bloomer formed a partnership with Thomas Davis and set up as ironmasters at Golds Hill. They then went on to lease the Pelsall Ironworks that had been established by Richard Fryer.

The Wesleyan faith provided the basis for business partnerships forged between the Bloomer, Davis, Frost and Thorneycroft families and were cemented together through marriage:

> Caleb Bloomer married Ann Davis, daughter of Thomas Davis, Timber Merchant
>
> Boaz Bloomer married Phoebe Davis, daughter of Thomas Davis, Timber Merchant
>
> George Bailey Bloomer married Charlotte Frost, daughter of Mathew Frost, Surveyor
>
> Samuel Frost (son of Matthew Frost) married Betsy Thorneycroft daughter of Edward Thorneycroft, Ironmaster

The Pelsall Estate was located beside the Wyrley & Essington Canal and comprised mines both to the north and south of the canal. Part of the land belonged to the Church of England and had been developed during the mid-1820s, first as a colliery. The furnaces were erected in around 1832 and the forges were completed in around 1834. Fryer had a vision of a large ironworks at Pelsall, served by a proposed canal that would connect the Birmingham Canal Navigations with the Trent & Mersey Canal, which would be known as the 'Birmingham, Walsall & Liverpool Junction Canal', first proposed in 1825. Richard Fryer was a banker who represented Wolverhampton as a MP (1832–1834). Despite the failure of the canal project, Fryer continued with the development of the Pelsall Ironworks. The forges were described in 1834 as containing puddling furnaces and mills capable of making 150 tons of sheets, hoops, plate and strip. The colliery comprised 70 acres and had a water (pumping) engine, three whimseys and five pits, with another two sinking. Richard Fryer died in 1846 and the estate passed to his son, William Fleeming Fryer, who passed on management of the operation to Bloomer and Davis.

Boaz Bloomer and his son, Boaz Bloomer junior, took a particular interest in the improvement of these works. They arranged for the reconstruction of the furnaces and ironworks. They also acquired additional mines on adjacent lands that belonged to the Charles family.

Henry Marten, Civil Engineer

The slim book, *Pits and Furnaces*, published in 1869, was written as a fictional account by Elizabeth Payne, but provides useful information about the period. Payne was also the sister of ironmaster Henry Marten and the wife of Alfred Payne, an industrial chemist. The account follows the travels of Mr and Mrs Hope and their children, Edward, Fanny and Harry, through the Black Country, drawing on true images of the period. It is a special description of the industries and people of the Ettingshall district, when Alfred Payne had a chemical works.

During Elizabeth Payne's time, the nearest smelting furnaces were at Bilston (Spring Vale), Millfields, Parkfields, Priestfields and Wolverhampton. There were many more which were placed near the line of the canal as it passed through Coseley, Bradley and Tipton. There are specific references in *Pits and Furnaces* to the Parkfield Furnaces, where Henry Marten was manager of the undertaking.

As local iron sources became depleted, the Parkfield Iron Co., like most of its neighbouring smelting furnaces, began to draw on supplies of ironstone from further afield. The stone was delivered to the canal basin at Spring Vale and taken up to the furnaces along their private railway. The foreign ore came from mines such as those at North Warwickshire, North Staffordshire and Cumberland. It was mixed with the local ores, still being drawn, to produce a type of iron which South Staffordshire became famous for.

The supply of coal tended to be less of a problem; as long as there was a cheap supply of coal to run the furnaces, the operation could proceed relatively profitably. However, to get the coal on their estate each ironmaster became more and more dependent on pumping operations to keep the mines dry. Water pumping had been carried out in mines for centuries. As a more sophisticated pumping plant was developed, deeper measures could be worked.

Elizabeth Payne, in *Pits and Furnaces*, marvels at God's work in creating faults and barriers within the coal strata that dammed up water and prevented it flowing throughout the whole of the coalfield. Such water was generally confined to specific underground areas called 'pounds'. Steam pumping engines were regularly employed to keep each of these pounds water free, although the actions of miners could inadvertently breach the boundaries, enabling water to flow elsewhere. At Parkfield there was a triangular-shaped underground pound which was drained by their Sandy Gay engine on the Ettingshall Lodge Colliery, as well as an engine owned by another ironworks. When this second engine was stopped, in about 1851, the Sandy Gay engine was not equal to the task of draining the pound. Slowly, the lower levels of ironstone and coal were drowned out.

The Parkfield Iron Co. then decided to dispose of its works and collieries; they were offered for sale by private contract in August 1852, with further advertisements in September and November 1852 and March 1853. The property then comprised 100 acres of freehold and 159 acres of leasehold mines. The works included four blast furnaces and two new 60HP condensing steam engines. A branch railway to the nearby Stour Valley Railway was also contemplated.

This was a time when the iron trade was good, yet it seems that the managing partnership which controlled Parkfield were having difficulty in finding a buyer. There were those who believed that whoever purchased Parkfield would find it a difficult task to make any profit. Edward Bagnall Dimmack eventually purchased the Parkfield concern for £27,000. He was then already extensively involved in the South Staffordshire and Monmouthshire iron trade. His partners in this venture were John Horton and S.H. Blackwell.

Edward Dimmack was born at Bilston in November 1808, the son of Jeremiah and Jane Dimmack. He married Anne Thompson on 17 October 1831 at Kingswinsford and also went into partnership with her brother, John Thompson, as iron merchants. They traded from Bradley Wharf,

The extent of Pelsall Ironworks, looking from Pelsall Works bridge to the railway bridge. (Source: Walsall Record Office)

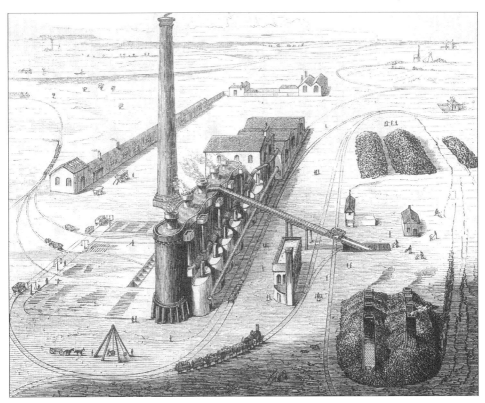

A view of Parkfield Furnaces, as reproduced for *Pits and Furnaces*, 1869. The drawing shows the railway system that served Parkfield, the incline to the top of the furnaces, the coke hearth, incline engine house, blast engine house and open pig beds.

Pothouse Bridge, along the canal to Manchester. They later joined forces with J.P. Firmstone of Highfields, who had an established iron merchants' business. This merger provided Dimmack & Thompson with a London office.

The Dimmack & Thompson partnership extended beyond the iron trade into the colliery trade, as Edward Dimmack and John Thompson were both associated with the management of the Sommerford Colliery at Willenhall and as Dimmack, Firmstone and Thompson were proprietors of the Bunkers Hill Colliery, near Bilston. Dimmack & Thompson also traded as iron merchants at High Bullen, Wednesbury, and were associated with the Gothersley Iron Co.

Edward Dimmack, Joseph Firmstone and John Thompson became involved with iron manufacture in South Wales. They took over the Pontypool Ironworks from C.H. Leigh in about 1850 and commenced working this establishment under the name of the Pontypool Iron Co. Iron smelting in the Pontypool area was a long-established industry; charcoal made from the local forests had been used to smelt local ironstones for generations. Coke smelting had eventually displaced charcoal at Pontypool and the works had been enlarged to four blast furnaces when the Pontypool Iron Co. acquired them. They also gained a share in the Coalbrook Vale Works at Blaina, Monmouthshire. Coalbrook Vale had two blast furnaces and had been previously worked by Brewer & Co. Apart from ironmaking, both Pontypool and Bilston had another trade in common, that of enamel manufacture. Pontypool enamels, like Bilston's, were internationally known. Edward Bagnall Dimmack did his best to improve the lot of his workers at Pontypool and provided a church and schools for their benefit. He was made High Sheriff of Monmouthshire for service to the community. Back home in Bilston, Dimmack had also risen in stature: he was appointed a Justice of the Peace.

Edward was thus an important and experienced local businessman. One can only feel that he negotiated a reduced price for the estate. Events would soon prove that this was actually a very astute purchase. Dimmack appointed Henry John Marten manager of the Parkfield Ironworks and Collieries. Horton's and Blackwell's direct association with Parkfield was brief. Other business commitments diverted their attention. S.H. Blackwell was particularly active, working in furnaces at Russells Hall and Bilston, as well as helping Fletcher and Solly to build new furnaces at Willenhall.

Henry Marten, who had married Dimmack's daughter, replaced both Horton and Blackwell when he became a junior partner in the Parkfield Iron Co. from 1855. Marten was a civil engineer by trade who specialised in hydraulic engineering. He had already been involved with several local contracts, including the Wolverhampton Waterworks schemes and Bilston sewerage. Henry retained his engineering practice aided by his brother, Edward Binton Marten, while simultaneously managing the Parkfield concern.

Meanwhile, the water problem had become a serious matter. Water from the Parkfield mines was flowing into the adjacent Millfields pound and then into Stow Heath pound. It threatened mining on a large scale. Fortunately, pumping operations were stepped up at Stow Heath and the problem was contained there.

Edward Dimmack became a director of the Bilston and South Staffordshire Coal and Iron District Building & Investment Society, which financed mortgages on housing properties. Another ironmaster, William Riley, was also a director of this society and the money they loaned assisted with the purchase of houses on the Bradley Field Estate, next to Pothouse Bridge.

Fortunately, the iron trade was in good shape and, therefore, allowed Dimmack and his partners to do well. Unfortunately, the history of the British iron trade has been dogged with cyclic periods of depression which are determined by the needs of supply and demand. During successful periods iron is over-produced and as stocks rise, prices fall. All sorts of factors can influence this process, but the pattern is always the same. Throughout the history of the industry there are periods of profit and slump. During 1855 another slump started to develop. The Crimean War had generated orders for iron goods, particularly cannons, rifles and shot, but as this war drew to a close the demand for iron began to diminish.

Ironmasters: Men of Property

The rewards from the iron trade provided social position and power for successful ironmasters. Profits accrued often financed a better lifestyle than hitherto experienced. The houses and homes of ironmasters often reflected the affluence attained.

Some served as Justice of the Peace, others in national or local government as Members of Parliament or local councillors. A few became Deputy Lieutenants for Staffordshire or Worcestershire. Notable names include James Foster, ironmaster and MP for Bridgenorth, who made Stourton Castle his home, while William Bennitt lived nearby at Stourton Hall. There was the Wolverhampton banker and ironmaster, a Justice of the Peace and Deputy Lieutenant for Staffordshire, William Fleeming Fryer, whose residence was the Wergs and Richard Smith, the capable agent for the Earl of Dudley who was a Justice of the Peace and Deputy Lieutenant for Stafford, Magistrate for Worcestershire and lived at the Priory, Dudley.

Henry Ward, son of Wolverhampton ironmaster William Ward, owner of the Priestfield Furnaces, was a Justice of the Peace and Deputy Lieutenant for Staffordshire and lived at Oaklands. Henry was related to the Bagnalls by marriage when he married Jane Bagnall, daughter of John Bagnall of West Bromwich. Several Wolverhampton ironmasters preferred the clean air of Tettenhall on the west side of the town.

Ironmasters also preferred certain parts of West Bromwich. Hill Top between West Bromwich and Wednesbury was the chosen place of residence for the sons of John Bagnall of Broseley,

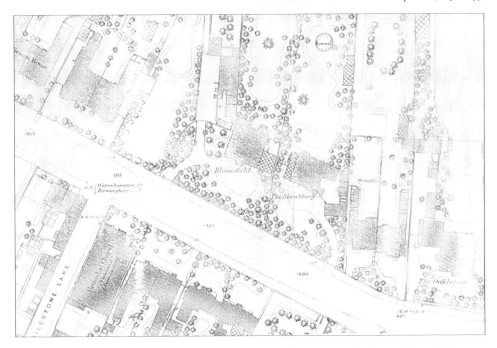

First Ordnance Survey map of Handsworth showing Bloomfield House, once the home of Joseph Hall, ironmaster, the person responsible for improving the iron-puddling process.

Modern view of Bloomfield House, Holyhead Road, former home of Joseph Hall.

Tomb of Joseph
Hall at Key Hill
Cemetery.

although Thomas Bagnall moved away from the district in later years. Charlemont Hall in West Bromwich was the home of the Price family of ironmasters.

Philip Williams, ironmaster of Wednesbury Oak, was another powerful individual who was both a Justice of the Peace and Deputy Lieutenant for Staffordshire and Worcestershire. Philip was High Sheriff of Staffordshire in 1857. He also served on the committee of the Birmingham Canal Navigations, a position that no doubt aided the carriage of his goods by canal. His position on the canal committee was terminated abruptly in December 1852 when a legal dispute developed between the canal company and his family over the drowning of mines at Birchills. Once this dispute was resolved, Philip regained a role with this canal company and at the time of his death in 1864 was deputy chairman. He was a director of the London & North Western Railway and for some fifteen years was chairman of the South Staffordshire Iron Trade. Philip Williams was unmarried. He lived in the family home at Wednesbury Oak, which was close to the heart of his industrial empire.

The management of Philip Williams & Sons was shared with three other brothers: Henry, Walter and William. All three were married with families that enabled the Williams ironmaster dynasty to continue. All three made Handsworth their home. Henry, who was a partner with Philip in a bank at Wednesbury, lived in Church Road. Walter had a home known as 'Woodville', beside the Holyhead Road and whose land bordered on the Austin Estate. William Williams lived at Oxhill, a large house placed on the hill near Oxhill Road and also near where St Augustine's church now stands.

Another Handsworth resident was Thomas Davies. He moved from Summerhill, Kingswinford, to the Austins, Handsworth, during 1838. The Austins was a three-storied brick mansion house, with grounds which were formerly the home of the Whateley family, notable Birmingham gun makers. The Austins Estate comprised the main house and farm lands. The Whateleys had improved the estate, making gardens with ornamental pools. A lodge guarded the entrance road from Sandwell Lane. Davies added another lodge on the entrance from Camp Lane. The Austins was let to Davies for a rent of £300 per year. It remained in the Davies family hands during the years they built up the Crookhay Ironworks at West Bromwich.

Joseph Hall resided for a time at Bloomfield House on the Holyhead Road. This was a large house that still stands, albeit extensively modified. William Orme, ironmaster of Stourbridge, also had a home on the Holyhead Road, as did William Henry Dawes.

CHAPTER 6

CONFLICT AND CHANGE
CHALLENGING TIMES IN THE TRADE, 1856–1866

There was yet another industrial revolution that happened throughout the British iron industry during the 1850s. This was marked by a period of industrial growth generated by the need for a greater variety of iron goods. The demands of railway rolling stock, structural ironwork, pipes for water and gas services as well as the continued demand for railway construction at home and abroad created new markets in this sphere. At the same time there were innovations with wire and tubes that generated even more business for the foundries, forges and mills.

Several smelting operations incorporated the ironmaking and iron-refining processes within their organisations while others chose to limit their production to one or the other. The number of firms that specialised in the conversion of pig to wrought iron increased considerably after 1850. Each firm had special brands, like the pig iron they were manufactured from, and they also took iron from a variety of sources. Local pig iron was mixed with other brands to achieve that special quality and uniqueness that was characteristic of the South Staffordshire iron manufacturers.

Pig iron was converted into finished iron through puddling, rolling and stamping. A whole range of different makes were produced; angle, hoop, horse shoe, nail rod, plate, rail, section, sheet, strip and tee. Manufacturers would engage in the production of any one or more of them. Finished iron was produced in a range of different qualities. Terms such as 'best', 'best best', 'treble best' and 'charcoal' were used to describe the various standards and had been in use for centuries. But as ironmaking techniques were honed through better engineering, more and more firms set up to work iron up to specific uses. Much of the iron used locally came from Staffordshire blast furnaces, but Derbyshire, Lancashire, Furness, Scottish, Welsh and Yorkshire irom was was also used in liberal quantities to obtain a special, or unique, quality.

The term 'forge' had come to encompass a broader meaning, including the buildings complete with puddling furnaces and mills, as well as the traditional use of stamping iron in the old type of finery and chafery. Foundries also had a wider application especially through the needs to supply engineering, metallic bedstead, fireproof safe makers, domestic appliances such as the kitchen range and structural ironwork. Foundries and forges increased in numbers, every one of them attempting to produce a better brand of iron than their competitors. In a lecture on the South Staffordshire industry, Daniel Jones recalled a nocturnal train journey through the district in 1854:

Travelling from Birmingham to Wolverhampton on a dark night about this period, one becomes conscious of being surrounded with a wild and brilliant scene – the country was apparently in uncontrolled flames; tongues of flames from 150 blast furnaces licked the sky, for then the economist had not completed the utilisation of waste gases. Two thousand puddling furnaces sent forth their flames, whilst hundreds of mill furnaces, cupolas, air furnaces and other flame generators, and thousands of pit-fires dotted over the country, illuminating [sic] the scene. Open coke fires burnt with sombre sullenness as compared with the brilliant brightness of the furnaces. The white light of the rows of street lamps helped to complete this amazing scene – a scene rendered more astounding from the constant changes and activity; the dampers of furnaces being alternately closed and raised, the opening of the furnace-doors and the trollying of the incandescent piles of rolls, or puddle balls to the squeezers or helves; red worms of bar-iron passing through the guide rolls, men stripped to the waist pouring down with perspiration moving to and fro, together with the whirling movements of the train through the district, screeching as it approached a signal-post as though it were escaping horror stricken from a prairie fire, each minute bringing a change of scene, and yet one continuous conflagration. Having reached his destination and collected again his wits, there could be but one impression left on the mind of the traveller who had witnessed the scene – that he found a veritable pandemonium, fit for Satan, 'gorgons and hydras, and chimeras dire'.

Daniel Jones was the grandson of George Jones, the ironmaster of Coseley, Bilston and Walsall. He had first-hand experience of the iron trade from an early age. His carefully written description has useful facts that show what ironworks were like. The buildings were very basic with minimal infrastructure. The mills, puddling and ball furnaces were laid out across the site, covered by a slate roof that was supported by cast-iron columns. Most structures were open at the side and those passing could often look in to observe the men and boys as they toiled.

Engine houses and furnaces were brick-built and the boilers often sat on a brick base, but there was little else of substance to surround the working place. An open design was essential because space was very important. The puddlers needed space to swing the balls of iron around to the hammers for shaping into bars. The labourers required space to move the reheated iron from the furnace to the guide rolls or mills. An even greater amount of space was needed to pass the red-hot iron through the rolls. Time and time again men would send the ever-increasing length of iron backwards and forward through the rollers until the required shape and thickness was attained. Ribbons of glowing metal snaked across the bed of the rolls only to be caught and turned back again by tongs held in the hands of the skilled workers.

Iron rolled into flat sheets was a product of many ironworks. The sheet iron produced was often corrugated and galvanised for roofing purposes. Some firms engaged in the manufacture of sheet iron and other produced corrugated and galvanised iron, sometimes both were manufactured. Galvanised and corrugated iron was a speciality of the ironworks in the Wolverhampton and Bilston districts.

It was during this time that the industrial heartland of diffuse districts East Worcestershire and South Staffordshire became known as the 'Black Country', a term which is cherished even to the present time. The person who coined the term, or gave it national credence, as previously mentioned, was the Mayor of Lichfield who used it on the opening of the South Staffordshire Railway.

By 1850 the whole Black Country was a patchwork of working mines; new mineral properties were becoming harder to come by. Iron, however, was in great demand and wise investment in the industry provided, at times, a good return on the capital invested. Some people chose to put their money into existing ventures, while others gambled for higher profits by establishing new concerns. The iron industry, at best, was a risky business as reverses in trade happened frequently. Investors usually lost out in these circumstances and pressed hard for the return of their money. Bankruptcies were common and the only people who profited were the solicitors and lawyers.

Those who entered the iron trade did so with caution. There was often a fine dividing line between success and failure. So many factors could influence the latter course. It helped if the ironmasters possessed coalmines, because then fuel was cheap. Large quantities of coal were required to smelt ironstone to make pig iron and then work it to a finished product. Although the working of mines presented additional problems it enabled their final product to be competitive in a very competitive market.

Fletcher, Solly & Urwick, Ironmasters

During the early 1850s the iron trade had come out of a recession and there was a heavy demand for iron. Coal and ironstone mines were opening as the Black Country iron trade increased production. It was the prelude to the most productive time for the Black Country trade. Inevitably, new investors saw opportunity for profit, including new entrepreneurs, Howard Fletcher and his brothers-in-law, Benjamin and William Urwick.

They purchased and leased land in Bloxwich and Willenhall with the intention of mining ironstone for local industry. It was the start of a venture that was to involve the Fletcher and Urwick family in the iron trade for a number of years.

Like many places in the Black Country, Willenhall had mines of coal and ironstone. The Thick Coal usually associated with mining in this region was absent in the Willenhall district, but there were seams of Bottom, Fire Clay and New Mine Coal. The ironstone measures comprised Blue Flats and Gubbin Ball and both were eagerly sought by ironmasters to smelt into pig iron. However, with the exception of the area around New Invention and Short Heath, much of the mining property around Willenhall was slow to be exploited.

Starting afresh required a special understanding of the iron industry. At Willenhall there was still land available and mines underneath that had not been exploited. Working these mines was, however, a gamble, because their extent was not known. Many coal and ironmasters have found to their cost that after sinking a mine the measures were broken and unworkable. The rewards could be great for a successful venture and no doubt a certain amount of luck would be needed.

By 1853 the ironmaster Samuel Holden Blackwell had become interested in the area. He already operated five blast furnaces at Russells Hall, near Dudley, which smelted local ores from an adjacent mineral estate. In 1852 he had also leased furnaces near the terminus Bilston Branch Canal from the Baldwin family. Two blast furnaces erected for John, Martin and William Baldwin and part of the Wallbutts Colliery were leased to S.H. Blackwell, according to an indenture dated 27 February 1852. Blackwell added three new furnaces to the plant, which became known as 'Bilston New Furnaces'.

Samuel Blackwell, like many other ironmasters, had limited sources of local ironstone at his disposal. He was constantly searching for new supplies, local or otherwise. The demands of his ten blast furnaces would inevitably use up local ironstone at a fast rate. Blackwell was one of the first, if not *the* first, to smelt Warwickshire ironstone. By 1850, Blackwell was boating ironstone from Bedworth to the Russell Hall Furnaces, and no doubt this trade was extended to Bilston New Furnaces when they were acquired in 1852. The Bilston Canal served the New Furnaces. It was a short branch that joined the Walsall Canal near Herberts Park. Mines were established at nearby Moxley and any stone raised there would have been boated along the canal to the furnaces.

Blackwell was an intelligent man with a passion for geology. After the Great Exhibition of 1851, he proved the worth of Northamptonshire ironstone and, almost single handedly, was responsible for ironstone quarrying being established there. It took a number of years for other ironmasters to accept the value of Northamptonshire ironstone, yet it did not deter Blackwell. He probably had this discovery in mind when he leased Bilston New Furnaces and extended the operation.

In March 1853, Samuel Blackwell purchased another untouched mineral estate at Willenhall from William Stokes. The ironstone mines had been proved on adjacent land and Blackwell would have thought it a worthwhile investment. The property comprised 40 acres of copyhold land between Somerford Brook and Little London known as the 'Harper's Estate'. There were other pieces of land adjacent to the Harper's Estate, untouched by mining operations, and there was opportunity to develop an area of at least 80 acres. Blackwell had many financial commitments at this time and chose to take partners in the Willenhall venture. He was joined by Howard Fletcher and then by his business partner, Benjamin Urwick.

It was quite a radical change for Fletcher and Urwick whose previous business experience was in the wine trade. Howard Fletcher was a successful wine merchant who lived at 8 Abblewell Street, Walsall. Benjamin Urwick was his junior partner whose home and business premises were at 33 High Street Birmingham. Both Fletcher and Urwick are listed as owners of the Birmingham property, which included a counting house and wine vault.

Howard Fletcher had been born in Walsall and was baptised at St Matthew's church on 11 December 1800. His father was Samuel Fletcher, a saddle ironmonger and respected member of the Walsall community. Howard was married twice. His first wife, Amelia Susannah, bore him two sons and a daughter, before her death in 1840. Henry Howard Fletcher was the eldest son, born in 1834 at the family home in Abblewell Street. Their daughter, Amelia, was the next born (1838) and then James Duppa (1840). Fletcher later married Ann Urwick from Weston, Radnorshire, and had another two sons and two daughters. It was a marriage that united the Urwick family with the Fletcher's and cemented useful business bonds.

Blackwell and Fletcher started to acquire other property adjacent to the Harper's Estate. About 15 acres located at Portobello, near the Somerford Brook (River Tame), the London & North Western Railway and the Wolverhampton and Willenhall turnpike were taken at the end of 1853. The land also included an area leased from the Willenhall Chapel Estates and an adjacent plot of copyhold land purchased from the estate of the late John Clemson. Both purchases provided the mineral property that was to be known as 'Portobello Bridge Colliery'.

Other property was either purchased or leased over a period of two years. Land between Little London and Sandbeds, which amounted to 28 acres, was leased from the Willenhall chapel, called the 'Stringes Estate'; about 4 acres of adjacent land was leased from Mrs Whitehouse and the Griffiths family; a further 2 acres of land called 'Dockey Pieces' was purchased from Samuel Hincks and 9 acres were purchased from John Clemson at Sandbeds. The church lands also formed an integral part of the Willenhall Furnaces Estate, they belonged to the Willenhall Chapel of Ease and sale of the lands had been authorised by Act of Parliament, 6 August 1844. The land had been advertised sale for private contract during 1853 in five lots. Most was virgin mineral property, only the 7-acre property known as 'Big Stone Piece' had been worked for coal. Here the top measures of coal 'had been recently gotten'. The money raised from the sales was used to provide a residence for the incumbent of Willenhall. Such was the revenue that it enabled two new churches, St Anne and St Stephen, to be built.

According to an indenture dated 28 November 1854, Blackwell and Fletcher paid £3,000 for the land on the Portobello Bridge Estate. Another agreement for the Stringes Estate involved the payment of £600 per annum. Blackwell, Fletcher and Urwick thus obtained a virtually unbroken corridor of land that extended from Portobello, through Little London to the Sandbeds. By the start of June 1854 about 85 acres of mineral land had been acquired and negotiations were in hand for more.

The partners were still undecided, however, as to which course of action to follow. There were at least two options open to them. One course, which would have suited Blackwell, was to sink shafts, erect engines and then sell the coal and ironstone to local furnaces and ironworks. The initial investment would be in plant and sinking, but charter masters could then be employed to get the coal and ironstone on their behalf. Another course was to mine the ironstone and then smelt the ironstone to make pig iron.

Benjamin Urwick wrote to his brother, William, in June 1854, to ask him if he was still interested in taking a cut in the business. Benjamin had been encouraged to believe that it was the best bit of mine left in South Staffordshire. It contained the finest quality of ironstone, the Blue Flats, which were then selling at 24s per ton. William Urwick decided to join them in the venture. William was a Ludlow solicitor and his keen legal mind was to prove very useful to the partnership.

Another person, Nathaniel Neal Solly, later joined this partnership. Nathaniel, who preferred to be known by his second name, Neal, had operated the Tividale Ironworks with his brother, but following bankruptcy had accepted a post as agent to the Welsh Slate Co. at Porthmadog. The Welsh Slate Co. operated the Rhiwbryfdir Slate Mines at Blaenau Ffestiniog. They sent slate along the Festiniog Railway to Porthmadog where Solly lived with his family. Neal Solly had been born in Clapton, Middlesex, and had toured Europe as young man, where he developed a keen

Nathaniel Neal Solly as seen in later life.

interest in painting and would have chosen this profession were it not for family commitments. By 1855 Neal Solly is also recorded as residing at Tettenhall Road, Wolverhampton. His move to the Midlands also accompanied the business expansion plans of his brother, James.

James and Richard Solly were proprietors of ironworks at Lea Brook, Tipton, by 1845. These works, which were located beside the Walsall Canal near B.W. Blades Brickworks, manufactured bar iron and also dealt in steel. Solly's factory was one of three separate and totally independent works that bore the name, Leabrook Ironworks. The other two were owned and operated at this time by Bagnall & Sons and the Chillington Iron Co.

On 1 June 1854, an informal partnership was therefore established between Samuel Holden Blackwell, ironmaster, Dudley; Howard Fletcher, wine merchant, Walsall; Neal Solly, ironmaster, Wolverhampton; Benjamin Urwick, wine merchant, Birmingham and William Urwick, solictor, Ludlow. No written agreement seems to have survived between these gentlemen and it was quite likely that none was drawn as they were still in the process of proving the mines. Simon Blackwell assisted the partnership from the start. His ground bailiff was sent to Willenhall in June to ascertain what the ground would yield.

Sinking appears to have started in several places on the estate. By August 1854 the coal measures had been proved on the Stringes Estate and the engines were ready for erection. The decision was then taken to erect blast furnaces and manufacture pig iron. At one pit, the sinkers had passed through the first coal measures, which were 6ft 3in thick. The ground bailiff sent Howard Fletcher the first cartload of coals, which he burnt in his house. Sinking was then suspended until the engine was finished. As work progressed through the winter, the whole works gradually came together. The site for the blast furnaces was selected on a piece of land known as 'Constable Dole' that was bounded by Stringes Lane and the Bentley Canal.

The responsibility of constructing blast furnaces fell to Mr Blackwell who arranged for three furnaces to be erected on the land. The site was compact and well laid out. The furnaces faced a new canal basin linked with the Bentley Canal. Colliery plants were erected at Stringes Lane and Back Lane, while iron-mining plants were erected on the Harpers and Portobello estates. These mines later became known as 'Portobello Bridge' and the 'New Plant'. A horse-drawn tramway was put down from the furnaces to the Stringes and Back Lane Collieries, which crossed St Anne's Road on the level. It then crossed Back Lane and the Temple Bar at Little London, before turning southwards to the New Plant and Portobello Bridge ironstone mines. Coal and ironstone was conveyed by the tramway to the new furnaces.

During 1855 Fletcher & Urwick purchased another mineral estate at Bloxwich from John Bealey. The mines, which amounted to about 24 acres, were located on the north side of Sneyd Lane and included a brick kiln and workmen's cottages. There were several pit shafts sunk to the coal measures and a tramway was in place to convey minerals to Sneyd Basin on the Wyrley & Essington Canal. By March 1855 the furnaces were almost ready. The ironstone mines at Portobello were also reaching a point at which mining could commence. Pits one, two and three had been sunk and gateroads were being opened out to get Poor Robins and Blue Flats Ironstone. £10,988 13s 9d had been spent on building the furnaces and associated plant. Each furnace was constructed of fire bricks encased in iron, which was then the latest development in furnace construction. In addition to the furnaces, two beam engines were erected. One supplied the blast for each furnace, the other was the incline engine which raised the charge to the top of the furnaces. Boilers also had to be installed to provide the necessary steam to work the engines.

The materials were provided by a variety of sources: Blackwell & Co. of Bilston supplied castings, the Corbyns Hall Co. provided iron plates and sheets and Solly Brothers, Leabrook, also sent some plates and rails. Firebricks were supplied by H. & B. Whitehouse, Coseley; Harris & Pearson, Amblecote and Harper & Moore, J. & W. Beddows, B. W. Blades and the Rose Hill Co. all supplied bricks.

Mining accounts tended to be regulated by the quarter days. That is Lady Day, Midsummer, Michaelmas and Christmas. It would seem that ironstone mining on Fletcher and Urwick's estate began in April 1855. Indeed, accounts for the first six months of 1855 show that ironstone was raised and sold on to the following firms for use in their own blast furnaces: William Baldwin & Sons, Bilston; William Bennitt, Oldbury; Chillington Iron Co. and Woodall & Smith, Windmill End. Coal was also supplied to a number of local people and firms, indicating that land sale had commenced at their pits.

S.H. Blackwell left the partnership when the furnaces were completed in 1855. The *London Gazette* carried the notice that the partnership between S.H. Blackwell and Fletcher, Solly & Urwick had been dissolved with effect from 16 June 1855.

Samuel Blackwell had an inquiring mind and was frequently involved with a number of projects at the same time. He seems to have constantly strived for a new mental challenge. Hence when he had solved the problems at Willenhall, he went on to something else. By 1855 he was assisting the establishment of the new South Staffordshire Waterworks Co. Although lauded by his contemporaries for his discovery of the Northamptonshire ores and his subsequent experiments with anthracite coals to smelt iron in the blast furnace, his business affairs were less than adequate. On 1 July 1855 Howard Fletcher, Neal Solly, Benjamin Urwick and William Urwick signed a formal contract of partnership. It was agreed that the period of partnership was to be for twenty-one years. Management of the firm was left in the hands of Neal Solly and Benjamin Urwick who received annual salaries of £600 and £300 respectively for their services.

A considerable financial investment was required. William Urwick had originally laid out £13,865 and Benjamin Urwick £4,808 19s for the furnace and mine development, but the payments made for the estates, at this time, had been more than anticipated. What Benjamin and William had contributed was adjusted and £7,773 6d was repaid to William Urwick with interest.

Thus Howard Fletcher held half of the finance capital, with Benjamin and William Urwick sharing the other half. Neal Solly's monetary commitment at this time was minimal.

Mining and furnace production began during the spring of 1855 using minerals from their estate. Good seams of coal were available locally and Fletcher, Solly and Urwick worked the New Mine and Bottom Coal for their purposes. Gubbin Balls and Blue Flats Ironstone was raised in their mines, but limestone was not available locally and was brought in by boat. Accounts for June 1855 mention that two suppliers of lime were John Brawn and George Strongitharm, who both had lime works on the Daw End Canal. Lime was a component of mortar but, although the works were under construction at this time, it is more likely that the lime mentioned was the first supply of limestone for the furnaces. The mines at Daw End had seams of thin and thick limestone. The upper thin seam was used for refractory, that is, ironmaking purposes, while the lower, thick seam was cut to provide blocks for building purposes.

With Fletcher, Solly and Urwick the decision to manufacture iron did not extend to refining it. Neal Solly had a close relationship with the Solly Brothers at Leabrook Ironworks and pig iron would have been passed on to Leabrook for working up to wrought iron. Within two years the Solly Brothers had added Great Bridge Ironworks to their operations and were working forty-three puddling furnaces on the two sites.

It was a feature of mining operations in the Willenhall district that as pits were sunk and developed, work commenced on new sinkings. Mining in these areas has to be viewed as a coalfield rather than a colliery. As the minerals were extracted in one part of a given area, new pits were sunk to get at the minerals in the other part. To work the minerals, Solly and Urwick would have employed chartermasters, one such person was Edward Bradley. Individual miners were paid by the chartermaster who recruited them for the manual labour of getting the stone and coal.

The methods of getting the two minerals also varied. With Blue Flats Ironstone there was a reasonably consistent bed of ironstone to be worked, while the Gubbin Balls had pieces of ironstone mixed with other materials. The seams were often narrow and miners sometimes worked within confined spaces. They worked through the seam, throwing the waste behind them to hold up the roof. For the coal miners, the seams were 4 or 6ft thick. Sometimes miners worked them on the rib and pillar system, others worked the seams by longwall methods. Miners working the shallow seams of coal were often referred to as 'Thin Coal miners' and their wages were usually less than those who mined the Thick Coal. Coal mining was carried on under more stringent rules than ironstone mining. Gas explosions and the risk of suffocation from 'choke damp' were ever present. Roof falls were also more common in the coal pit. Miners' lives were frequently at risk due to these problems.

The butty, or chartermaster as the job was also known, was usually paid on a royalty basis for the amount raised. His actual role varied from employer to employer. With Solly and Urwick, it seems they retained a strong involvement in the management of the mines and men. An enduring feature of the chartermaster mode of working was the truck system, where wages were paid partly as goods from the company shop or the chartermaster's beer house. Despite laws that regulated the practice, it was still prevalent in Willenhall in the 1850s, as was pointed out by the *Mining Journal* in January 1859:

> A meeting of colliers to protest against the Truck System was held on Monday evening near Willenhall, but the agitation appears to be quite unorganised at present. If the question were taken up unitedly by the colliers, there can be no doubt but that the system would be put down. The worst form of the evil is the Chartermasters keeping a public house and getting their men to go and drink there. It is a known fact that a charter master, or butty, who keeps a public house will contract to get coals at a much cheaper rate than one who does not and hence it is difficult for masters to put down the system. The results are very bad. The butties are induced to employ unmarried young men in preference to those who have families, as they spend more on drink. In combating such an evil the colliers would have the sympathy of most of the masters, and of all the right thinking part of the community. Many attempts to put down the practice have been made, but in a few instances they have been preserved in.

Despite a degree of adverse public opinion, many local ironmasters chose to have shops for their workers, with some taking pride in the way they were operated. William Ward had a shop at Priestfield. Speaking to the Midland Mining Commission in 1843, Ward defended his shop by stating it provided good value. His stocks were always purchased with cash, which enabled him to pass on some of the benefits to his workers. These company shops made money for the owners, but not all had the scrupulous virtues of William Ward. It is not surprising to find that Fletcher, Solly and Urwick had a shop at Portobello, known as the 'Albion Shop'.

In February 1859, Edward Bradley was prosecuted under the Truck Act. The mother of one of Mr Bradley's workers went for his wages and was told that unless she spent some of the money in the Fletcher, Solly & Co. shop her son would have no more work. On going to the shop, the people there insisted that she spent all the wages in the shop. When she decided to spend her son's wages elsewhere, he lost his job. Another two people also came forward with the same complaint and it was further said that Bradley had been fined for the same offence before. The magistrates decided to fine Mr Bradley, but this amounted to £10 and no doubt was of little deterrent. It is also evidence of the power wielded by the chartermaster. Prosecution under the Truck Act, at this time, was effectively an occupational hazard and through their status as employers, working conditions could be made harsh for those who did not accept the system. Few would dare to cross them, unless they chose to look for employment elsewhere. The shop undoubtedly provided a lucrative source of revenue and the spending of wages there was also seen to prevent drunkenness amongst the miners, particularly the family men. Horatio Samuel Fletcher said as much when interviewed by the government inspector in 1842 about the truck system. For the family man it was a way of securing food for his wife and children.

The underground workings might have been the preserve of the chartermaster and his assistants; the surface tended to be under the charge of the owners or an agent. Minerals were brought to the pit bank for disposal by land sale or transport elsewhere. At mines such as those owned by Fletcher, Solly & Urwick, the bulk of their traffic was by a narrow-gauge tramroad to the Willenhall Furnaces. It was a feature of the mines in this district to use steam engines to raise the minerals and to let the men into and out of the pit. The horse gin might also be used, but the steam engine was far more common.

On the pit bank the banksman and his assistants would assist with bringing the minerals out of the mine. The engineer would work the engine and often feed the boiler with slack. The Little London coal and ironstone field comprised of pits worked by different companies. There were a number of engines dotted across the land from Noose Lane to the Temple Bar, each working a group of pits at the same time. It was a common sight to see the flat three-link chain strung out from the engine house in straight lines to the pit shaft headgear, the course guided by pulleys fixed to wooden posts. The economy of using one engine made sense, unfortunately, in practice, the working of the system was none too safe. For example, the Osier Bed Co. worked a number of shafts on either side of the New Plant Colliery. In January 1859 there was an accident on their part of the coalfield. A single engine, under the control of one man, worked four pits. The posts and the pulleys were not always kept in the best of repair; when the engine was set to work the chain was forced from one pulley wheel and snapped. Three men fell to their death as a result. Mr Sparrow's ground bailiff, James Williams, had requested the carpenter to repair the pulley and post before the accident, but nothing had been done.

By the time the census was published in 1861, mining in Willenhall had become an important employer. The *Mining Journal* remarked that the population of the parish had increased by 45 per cent since the 1851 census and attributed this to the influx of miners. Bilston, by comparison, had shown only 3 per cent increase for the same period. It was concluded that mining in Bilston was on the decline as the mines became exhausted.

The iron trade was good when Fletcher, Solly & Urwick started their furnaces. The Crimean War had generated many government contracts for munitions and other sundry ironwork. Both pig and puddled iron commanded high prices in the market place. Finance for the venture was provided by investors and mortgages from the bank. William Urwick provided an important part of the capital that amounted to £12,000. Another part of the necessary capital was funded from mortgages. By December 1858 the following mortgages had raised significant capital for the working of the furnaces:

Portobello Bridge Colliery	Samuel Perks & Edward Shaw	£7,000
Stringes Colliery and	Edward Coates, J. Southern,	
Furnaces, freehold land	R. Burton and T. Penson	£8,000
New Plant Colliery	Wolverhampton and	
	Staffordshire Bank	£12,790
Sneyd Lane Colliery	Richard Urwick	£2,000
Total Mortgages		**£29,790**

The new property arrangements led to a revision of the shares held by the partners. The property was now divided between the four partners in 302 equal shares: Howard Fletcher (109 shares), Nathaniel Solly (45 shares), Benjamin Urwick (74 shares) and William Urwick (74 shares).

Willenhall Furnaces were located beside the Bentley Canal at a place known as 'Sandbeds'. A basin was made into the furnace yard. There were eventually three furnaces here, each 45ft high and 13ft at the boshes. Local coal and ironstone provided a major proportion of the minerals used to make pig iron. Coal pits at Little London and Portobello raised coal and ironstone for local consumption. In 1870 there were five at work, all capable of raising 25 tons of coal per day. Robbin, Gubbin, Ball and Blue Flat Ironstone was also taken from the mines. From 1855 until 1862 the minerals were taken to the furnaces along horse-drawn tramways. From 1862 steam locomotives

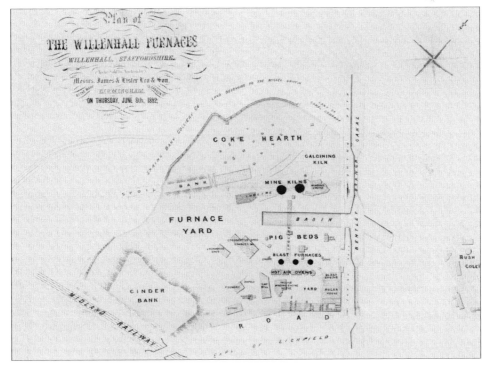

Willenhall Furnaces sales plan, 1882. The plan reflects the changed arrangement of the furnace layout. There was an incline to the top of the furnaces, an open-air pig yard and calcining kilns where the ironstone was heated with coal to increase the iron content in the material that went into the furnace mix. (Source: Lee Crowder MS, Birmingham Archives)

replaced horses on many haulage duties along the tramways. Two thirds coal and one third coke was mixed with calcined ironstone and lime for smelting. Local coals included Bottom and New Mine. The Bottom Coal was used to make the coke, while New Mine Coal was used in the furnace without being coked. Other coke and limestone was brought by canal. Iron ore from other districts was also brought to the furnaces by boat.

Crisis in the South Staffordshire Iron Trade

Masters and men benefited from the demand of iron generated by the building of new railways at home and abroad. But as demand slackened one local railway contractor, Fox, Henderson & Co. failed. Their main works were located at Smethwick and were known as the 'London Works'. They specialised in the making of railway rolling stock and the supply of track and structural ironwork. Their reputation received a great boost through the supply of the structure of the Crystal Palace for the Great Exhibition in London. Severe losses caused through the making of the Zealand Railway in Denmark led Sir Charles Fox and his partner, John Henderson, to suspend payments in 1856. Litigation followed concerning shares in docks in Paris. Fox and Henderson were compelled to submit to bankruptcy proceedings with the result that their estate and business was split up and sold off. The lengthy process took several years to resolve. In the meantime, their failure proved to be the herald of others that were to follow.

The iron trade also faced a commercial setback during 1855, but with support from local banks, firms in difficulty were able to continue in the trade. Iron production in South Staffordshire had actually peaked by 1857. Blast furnace statistics for September 1857 demonstrate that out of 180 furnaces in that region, 155 were at work producing iron. Within a month there was a sudden downturn in trade and several firms were caught out. The price of iron fell dramatically. Several established firms faced bankruptcy, including Fletcher, Rose & Co.; Thomas Morris & Son and Frederick Charles Perry.

Blast furnaces were blown out and stocks were devalued over night. For firms such as Fletcher, Solly & Co. the prospects were not good. They had a wage bill to meet for their employees. Debts had to be paid for coal, ironstone and limestone purchased elsewhere. But most importantly, annual and quarterly payments had to be met for the leases, royalties and land purchases. Amounts such as these could cripple a firm when sales dropped off. Fortunately for Fletcher, Solly & Co., other sources of income were possible. Mortgages of their freehold property could be arranged and both Fletcher and Urwick's firms had additional capital to invest if needed. Other firms were not so lucky.

The Birmingham Bankruptcy Court was inundated with petitions for bankruptcy. Amongst the failures was Solly Brothers of Leabrook and Great Bridge. They suspended payments in November 1857, but after an examination of their affairs were allowed to continue under inspection. Fletcher, Solly and Urwick were conscious of the fact they still owed money for various land purchases and resolved to simplify matters regarding their property agreements. It was agreed in 1858 to pay off the outstanding property debts out of the capital and other partnership funds.

One of the largest failures in 1857 was that of William Riley senior and William Tomkinson Riley, who traded as William Riley & Sons at Millfields Furnaces and Bentley Furnaces. William Riley had taken over Millfields Furnaces and associated mines from the widow of John Walker, the previous owner, and had also added the Bentley Estate that was leased from the Countess of Lichfield. Their liabilities amounted to over £240,000, but only had assets worth £80,000. The firm had been insolvent in 1855, but had still continued to trade and built up even greater debts.

Trade was nearly at a standstill during November. The Chillington Co. discharged 300 men; Lord Dudley's agent shut down furnaces, closed pits and reduced wages; Thomas and Isaac Badger put out their Old Hill Furnaces; William Haden shut down his furnace at Dixons Green and Woodall & Smith of Windmill End put out their furnaces. Cochrane's of Brierley Hill discharged about a hundred workers; Simon Holden Blackwell reduced wages; New British Iron Co. closed down three furnaces and John Jones blew out three of the five furnaces at Birchills and closed eleven pits. Creditors' meetings were also held for Motteram and Deeley of Toll End Works, which had failed in 1855 but had continued working, Rose, Higgins & Rose (Bradley and New Bradley works), Brayford & Lancaster (Birchills Ironworks and Pleck Ironworks at Walsall and Staffordshire Ironworks, Greets Green), and Poole & Co. of Wolverhampton Furnaces. In January 1858 the *Birmingham Journal* published a list of twenty-five firms in trouble:

	Name of firm	*Liabilities (£)*	*Estimated assets (£)*
1	W. Riley & Son	240,332	80,000
2	F.C. Perry	85,527	32,600
3	Rose, Higgins & Rose	72,500	29,417
4	Solly Brothers	75,585	75,551
5	Fletcher & Rose	40,466	21,322
6	Woodall & Smith	69,627	28,000
7	Brayford & Lancaster	40,000	8,000
8	Wright & North	39,104	20,231
9	T. Morris & Son	40,000	12,609
10	Pearson & Kenrick	38,000	14,000
11	Motteram & Deeley	59,562	12,596
12	Stone & Son	20,000	15,000
13	G. Bate & Son	13,000	300
14	E.T. Wright	12,000	12,000
15	Dr Mannix	10,000	2,500
16	Fletcher Brothers & Green	13,148	8,036
17	Richard & Son	10,000	2,500
18	W.H. Gregory	8,000	4,000
19	T.H. Pemberton	33,000	25,000
20	Bradley Hall Co.	11,964	659
21	Hodgetts & Son		
22	T.H. Giles & Co.	30,000	46,000
23	Joseph Spencer	6,909	2,800
24	Samuel Griffiths	150,000	Not known
25	Wolverhampton Iron Co. (E.T. Poole)	15,197	5,000

A few firms, such as Solly Brothers, survived the creditors' meetings and were allowed to continue trading. It was crucial that this sanction was gained. Those who flaunted their creditors did so at their peril, as James Thompson discovered. Thompson was the senior partner in the firm of the Bradley Hall Co. which controlled the Bradley Hall Ironworks. When questioned about the serious discrepancies between assets and liabilities, James Thompson made the comment 'if you know nothing about a forge, you had better say nothing, for I'll go out and not be bothered with you.' He left the meeting and the creditors unanimously voted to have the firm wound up at the bankruptcy court.

Thomas Davies & Sons fared badly during a slump in the iron trade. Prior to 1855 they ran profitable operations that included the Crookhay Furnaces, Colliery and Ironworks, Bull Pleck Colliery, Neachells Colliery and part of the Spon Lane Colliery, all of which had been built up during the 1840s. Most of the securities were given over to Dudley & West Bromwich Bank during 1855, during a period of poor trading. The bank hoped to support the firm, but with the continued poor trade did little to prevent the bankruptcy of Thomas Davies senior and Thomas Davies junior during 1857.

Davies' ironmaking activities comprised four blast furnaces. Their ironworks buildings were covered by wood and slate roofing in the local style, open at the sides and supported by columns. The five largest buildings were each 167ft long by 42ft span. The space was needed for the various

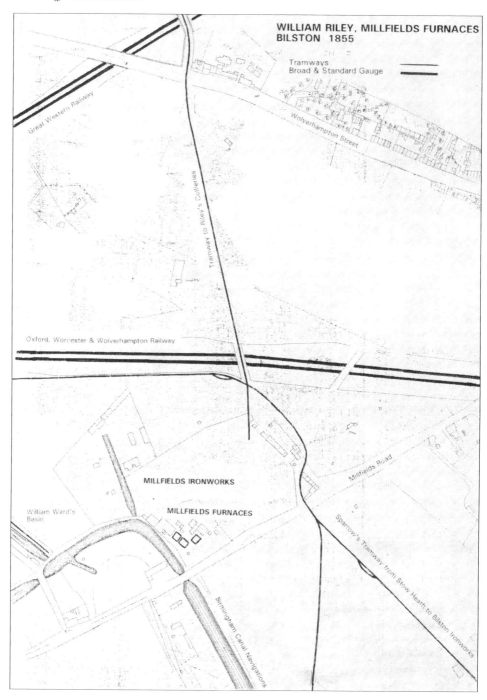

WILLIAM RILEY, MILLFIELDS FURNACES BILSTON 1855

Tramways
Broad & Standard Gauge

Great Western Railway

Wolverhampton Street

Tramway to Riley's Collieries

Oxford, Worcester & Wolverhampton Railway

MILLFIELDS IRONWORKS

MILLFIELDS FURNACES

William Ward's Basin

Millfields Road

Sparrow's Tramway from Stow Heath to Bilston Ironworks

Birmingham Canal Navigations

William Riley's Millfield Furnaces, *c.*1855. These furnaces, like many others in this part, had long tramways that brought both ironstone and coal to the furnace yard.

puddling, mill and heating furnaces which totalled seventy-two in number. There were also the steam boilers that were heated by the furnaces. Space was also needed for the rolling mills where the puddled iron bars were passed back and forth. The arrangement of rolls became commonly known as a 'train', perhaps as a result of railways spreading across the countryside. Sales details for the Crookhay Ironworks in 1857 mentioned an 8in guide train, a 9in merchant train and 16in merchant train, a boilerplate train and two sheet-iron trains.

Both the furnaces and ironworks were taken over by George Thompson & Co., who quickly developed financial difficulties. George Thompson had been employed by the New British Iron Co. where he had control of all South Staffordshire operations and a salary of £1,000. When Davies & Son failed he decided to leave Corngreaves and run his own business, taking over the Crookhay Works. Unfortunately the gamble failed, accumulated bad debts proved his undoing. By 1863, Crookhay Furnaces and Ironworks were in the hands of H.O. Firmstone. The ironworks and furnace site built up by Davies was now much reduced. Thompson, and later Firmstone, operated only seventeen puddling furnaces and a ball furnace. There was a train of 18in for boiler-plate rolls, a train of 20in for forge rolls and a train of 16in for merchant and sheet rolls.

Solomon Woodhall and Josiah Smith were simply extremely unlucky. They had three blast furnaces at Windmill End, ironworks, collieries and a boiler yard in their possession leased from Sir Horace St Paul, Baronet. They had spent a large amount of money draining the mines and had freed many of the mines from water. Coal and ironstone measures remained available for acquisition. Unfortunately, Woodhall and Smith's money ran out before they could exploit these coal and ironstone measures, therefore, others were to profit from their investment. Woodhall retained some property and continued to operate a foundry and boiler yard.

The slump had a serious affect on the Edward Bagnall Dimmack partnerships. Following the collapse of their London establishment with debts amounting to £100,000, Dimmack, Firmstone and Thompson sought to realise this deficit through disposal of their South Wales holdings and the Bunker Hill Colliery. Financing this became an increasingly difficult problem as property values were falling. It is perhaps a tribute to Dimmack, Firmstone and Thompson that they were able to raise such an amount in difficult times. Following the disposal of these estates, Edward Dimmack concentrated his efforts on the Parkfield concern. Henry Marten, his able son-in-law, was seen as the potential saviour of the business. Regrettably, the financial affairs of Dimmack's other partnerships dragged the Parkfield Co. into the mire. The resulting bankruptcies of Dimmack and Marten left the Parkfield Works in the hands of the Wolverhampton and Staffordshire Bank, with mortgage debts also owing to the Shifnal Bank. Charles Shaw and William Hatton headed the deputation of creditors to champion the Parkfield cause.

The continued survival of the Parkfield concern must be attributed to the skill of Henry Marten, who was able to contain the water and drain the mines. The *Mining Journal* for 1857 describes the events which led to the minerals being reclaimed:

The Iron and Coal Trades of Staffordshire. September 1857

The natural consequence of this state of things was the value of the Parkfield estate was greatly deteriorated; operations being necessarily confined to the minerals lying within 50 or 60 yards of the surface. It was in this state when purchased by the present proprietors, Messr Dimmack and Marten; and although it was by very many regarded as perfectly hopeless task, they determined, after careful calculation, and acting upon the advice of men of experience, in whom they confided, to make an effort to recover the valuable minerals which were submerged. So convinced was the Millfield Company that the drainage of the Parkfield pound was hopeless, that they actually removed their engine. Operations, however, were commenced by the Parkfield Company for the purpose of draining their pound. The power of their engine was augmented three fold by increasing the diameter of the pumps from 13 to 16 in., by lengthening the stroke, and by so altering the engine as to enable it to work 11 or 12 instead of 6 or 7 strokes each minute. After some time spent in preliminary operations, pumping was commenced in April last year, and in a few weeks the water was so far lowered that the new mine coal and the getting rock ironstone was drained. Several difficulties, however, interposed to prevent the realisation of the immediate success which this first result promised. The gradual removal of the old lift, pipe by pipe, in a narrow shaft, and replacing them by new lengths, was a very difficult and tedious operation. In addition to this, it was found that the water had during the years it had lain in the pound, absorbed a considerable quantity of freed sulphuric acid, and other impurities, which rapidly corroded the iron of the pumps and apparatus, and it was found necessary, in consequence to substitute three brass barrels weight upwards of a ton each. A new and very ingenious method of balancing rods, by means of a chain and wheel instead of a balance tip was adopted, and this proved much cheaper, more effective, and far less cumbersome than the original plan. Up to the end of July last, from the completion of all these improvements, the quantity of water raised averaged 1,200,000 gallons, or sufficient to fill a pond 1½ acres in extent and a yard deep, each day. The water was pumped into the canal, and entering it at the highest level was of considerable value to the Canal Company, sufficient being supplied to fill the locks about twenty eight times a day. The water has now been drained to the bottom of the shaft, a depth of 160 yards, and the roadways and workings are being cleared preparatory to re-working the measures which have been long under water. These operations have directly drained an area of nearly 1000 acres in extent; but in addition to this, the proprietors of the Millfields Collieries are preparing to replace their pumping engines, so as to re-open their mines, now that water ceased to flow from the Parkfield pound into theirs, and the proprietors of the Stow Heath mines will be able greatly to reduce their expenditure in pumping. As an illustration of the extent to which the value of mining property will be enhanced by the result effected by these operations, it may be observed that the minerals belonging to the Parkfield Company alone which will be rendered available are expected to yield a supply sufficient to keep their furnaces in operation for an additional period of 20 years beyond the time which consumption of the upper measures would have taken – in other words, the success of this spirited enterprise will enable the proprietors of that company to produce 500 tons of pig iron per week for this period, in addition to the yield of the upper measures.

The draining of the mines enabled the Parkfield Co. to increase iron production. Larger smelting furnaces were constructed between 1858 and 1859, increasing the iron production at a time when the iron trade was in a state of uncertainty.

During September and October 1859, the business affairs of Edward Dimmack suffered a serious setback. Although the Parkfield concern was now considered a profitable venture, some of his other business ventures proved to be a drain on his resources. Losses in these schemes meant that Dimmack was forced to apply for bankruptcy, with protection under the private arrangement clause, in September 1859. Dimmack's bankruptcy led to Henry Marten also making an application the following week. The complete business affairs of the Parkfield Co. were suspended and receivers, Horton and Shaw, were appointed. The affairs at Parkfield were again threatened, but fortunately the works were allowed to continue.

In times such as these mortgages were classed as debts and the banks, which provided the mortgages, became major creditors. The Wolverhampton and Staffordshire Bank had loaned large sums against the mineral estates and a smaller amount had also been loaned by the Shropshire Bank. In a hard-fought deal between the banks and receivers, the Parkfield Works was left in charge of Henry Marten. Edward Dimmack was 'retired' as senior partner and his other business affairs still had to be resolved.

The problems that happened at Parkfield were reported widely. The *Wolverhampton Chronicle*, *Birmingham Gazette, Birmingham Journal, Mining Journal* and *The Engineer* all printed articles concerning Parkfield. It was, perhaps, *The Engineer* who was most sympathetic to the Parkfield cause. There is clear criticism of the joint-stock banks whom they saw as willing lenders of funds as long as they could receive the sizeable interest such transactions produced. When a firm was in difficulty, the banks were amongst the first to close it down and realise its assets.

Dimmack & Marten came very close to losing Parkfield. The Wolverhampton and Staffordshire Bank was in no mood for leniency. It was only through the tenacity of Horton and Shaw that a deal was finally agreed. The compromise solution was to leave the profitable Ettingshall Lodge Colliery in the hands of the bank, who were then at liberty to let it to other mining interests.

Mr Hatton proposed that the Wolverhampton and Shifnal Banks were to be considered joint landlords and let Mr Marten keep possession of the works, paying £1,000 a year rent for the works and plant as well as royalties of 1s per ton ironstone, 9d per ton coal and 8d per ton slack. The minimum rent was to be £3,000 per year and from the divided four-fifths was to be paid to the Wolverhampton Bank and one-fifth paid to the Shifnal Bank. The Wolverhampton and Staffordshire Bank was unwilling to accept this proposal and had doubts about Marten's ability to carry the works on. Hatton and Shaw were confident that Marten's friends would enable him to carry on the business. Unfortunately, matters were swayed by the claim for rent and it was especially worrying as it was delivered prior to the creditors' meeting on the following Monday. This brought matters to a crisis and the case went before the courts.

Birmingham Bankruptcy Court was filled with a large number of creditors who had arrived for the first meeting. The atmosphere must have been tense. The court sat behind closed doors and reporters were excluded. Findlay Knight, who appeared for the petitioners, drew the judge's attention to the rejection of the plan proposed to save Parkfield. After a lengthy discussion, and despite objections from solicitors acting for the bank, it was decided that Dimmack and Marten needed more time and an adjournment was granted and a new date was set for 2 November 1859. Thus Dimmack & Marten gained more time, which was enough to thrash out a compromise deal with the Wolverhampton and Staffordshire Bank. When the adjourned meeting was reconvened on the Wednesday a modified proposal was presented to the creditors. On this occasion reporters were admitted to the court and a summary of the proceedings was published later by the *Birmingham Journal*:

> Birmingham Journal, 5 November 1859
>
> We may here state that terms having been offered by which the Wolverhampton & Staffordshire Banking has been satisfied, Messr Dimmacks and Marten's proposal to pay their unsecured creditors 20s in the pound, by quarterly instalments of 1s each, to commence on the 1st March, have been accepted, and only await confirmation of the next meeting. The Wolverhampton and Staffordshire Banking Company have accepted that which they regarded as a satisfactory guarantee for the payment of any deficiency that may arise from the sale of the Ettingshall Colliery; and a sum of £20,000, the amount agreed to be accepted, by the bank in full discharge of one debt of £45,000, this being in addition to the first mortgage of £16,000, secured on the Parkfield Estate. The Shropshire Bank has also been arranged with; Mr Marten will carry on the Parkfield Works, Mr Dimmack, we are informed, retiring, until his affairs in the partnership with Mr Thompson have been arranged.

The whole contest was clearly a hard-fought battle and provides suitable illustration of the problems ironmasters faced in their day-to-day business lives. Yet the outcome was the salvation of the Parkfield Co.; it continued to work the mines and furnaces for another fifteen years. Employment was not only provided for miners and ironworkers, but also the host of allied industries which supplied, or purchased from, the Parkfield Co.

Edward Dimmack now lived at Penn Fields. During his remaining years he still kept close ties with the Parkfield Co., served on the bench of the Sedgley Magistrates Court and also served a term as Deputy Lieutenant of Staffordshire. Magistrates courts dealt with the many misdemeanours of public life which ranged from evasion of the toll at the turnpike gate, through to stealing and cases of assault. One such case Dimmack presided over was the claim of assault in a local beer house; the complainant said he had been hit about the head and had spent time in hospital. On cross examination it transpired that his wife had left him some nine weeks earlier as he had had an affair with the defendant's wife. The case was dismissed because nobody was willing to confirm what had happened. Perhaps it was no small wonder, because the defendant was William Perry, better known as the 'Tipton Slasher', the former English boxing champion.

Samuel Holden Blackwell was less fortunate with his business affairs. When Blackwell died, in 1868, he was remembered more for his business failures than his successes with ironmaking and geology. Samuel Blackwell's business empire folded in 1860. Payments were stopped in February 1861 and the Bilston New Furnaces & Collieries were advertised for sale in July 1861. The Dudley Branch of the Birmingham Bank handled Blackwell's business affairs. His failure left large debts that took years to sort out and eventually led to some £130,000 being written off as a bad debt. This, and dealings with other ironmasters' debts, was a major contributing factor to the fall of the Birmingham Bank in 1868. The Russells Hall Estate was sold off to pay Blackwell's bankruptcy debts. The Bilston Furnaces were returned to the Baldwins, but remained out of use until final demolition.

Blackwell, during his remaining years, was involved with furnace management and with his death the trade lost one of its most able sons. Samuel Blackwell made some valuable contributions that included the development of the Seend Ironworks in Wiltshire, on the Kennett & Avon Canal. Two blast furnaces were erected there in 1860 to exploit a large reserve of ironstone, using the canal to bring in coal and limestone and carry out iron to the Port of Bristol. Fellow South Staffordshire ironmaster, William Gibbons, was appointed a director of the company formed to work the concern in 1861.

Another of Blackwell's achievements was the use of anthracite coals for ironmaking in South Wales' furnaces. All these counted for little, however, at the time of his death. The *Mining Journal* summed up his life in a brief obituary, laying the cause of his troubles as his constant enquiring nature, flitting from project to project and not taking more care with the responsibilities of management. With his failure many were put out of work. However, his associates, such as Edward Jones, furnace manager, were able to take their skills forward into other jobs in the trade.

John Howard Blackwell also suffered business failure. He had been in partnership with George Bennett at the Smethwick Ironworks, until bankruptcy in 1856. Court proceedings settled payments in 1858, but these debts and payments continued to be made in small amounts through to 1870. Meanwhile, in 1861, J.H. Blackwell accepted a post of manager for the Nerbudda Iron Co. in India.

The Hickmans, George and Alfred, encountered a difficult trading period as debts accumulated. The crisis of 1855 had affected their trade. They then operated the Bilston Brook Furnaces, Stonefield Furnaces and Ironworks, Groveland Ironworks and one of the Leabrook Ironworks with various degrees of success. It was an extensive trade and perhaps one that was overstretched for the time. They made a private arrangement with creditors to pay off their debts in instalments, as much as 10s in the pound, and some £40,000-£50,000 debt was paid off by November 1856. Unfortunately, they then took on the might of Philip Williams over a comment made about their solvency. The court case for slander presented before Stafford Court did little to help their business. George and Alfred succumbed to the bankruptcy courts in 1860, where proceedings found them lacking in the treatment of two people, John Barber and William Wooley. Barber acted as merchant for the sale of their iron and Wooley was a boiler maker. Wooley was particularly hard done by and filed his petition for bankruptcy some two weeks after Hickman.

These troubled times heralded a number of changes in ownership. Wolverhampton Iron Co., operated the Wolverhampton Furnaces, was established in the mid-1820s. Following Poole's bankruptcy, ownership passed to Isaiah Aston, in partnership with Corns, which lasted until 1860 when Aston took a new partner, Richard Edward Shaw.

Change also occurred at Gospel Oak when the long standing partnership of Messrs J. and E. Walker was replaced by Walter Robinson, who took over the operation of the rolling mills there under the style of Walter Robinson & Co.

The firm of W. Baldwin & Co., Bovereaux Works, bankers and tinplate manufacturers, disposed of their tinplate works to Stephen Thompson and Mr Hatton in 1860. The change was planned after the death of William Baldwin, who took a very active part in management of the concern. Mr Martin Baldwin, the only surviving partner, decided to reduce his commitment to the operation of the furnaces alone.

Walter Williams (1795–1867) was a member of the wealthy Williams family that were associated with various mining ventures, as well as the Birchills, Union and Wednesbury Oak Furnaces. Walter had built a successful ironworks business in West Bromwich. Born in Wednesbury, in later life he made his home at Woodville, a house that lay alongside the Holyhead Road, in Handsworth. During 1860, in his sixty-fifth year, Walter decided to retire from the ironmaking trade and gave up his various works at the Albion and Greets Green, which lined the Walsall Canal. The Albion Works were divided up into separate ownerships (Albion and Britannia), with Samuel Griffith gaining control of Britannia, while his works at Greet Green (Staffordshire) were also taken over by Samuel Griffiths, iron merchant of Wolverhampton.

The legacy of the crisis of 1855 and crash of 1857 was a thread of bad debts that was passed from ironmaster to trader and back again. New ventures were particularly vulnerable, but there was still a reservoir of new investors willing to take the chance. The ironworks on the Cape Arm frequently changed hands at this time. These included the various Cape and Grove works whose origins can be traced to the time of the nearby Fox, Henderson & Co., whose demand for iron was a factor in their establishment.

Samuel Griffiths was particularly active at this time, gathering up iron furnaces and ironworks in a brief frenzy at a time when prices were low. He was a predatory and cunning soul, which *The Engineer* in 1862 came to describe as 'notorious'. Griffith's house of cards collapsed about him in 1862, however. Born in Bilston and christened at St Leonards, Bilston, on 9 January 1814, Samuel Griffith never lived a dull life. He had interests in various trades and often faced bankruptcy, but was clever enough to frequently escape its consequences, often settling with creditors by taking money from one source to pay another in a fine juggling act. Arrest and periods in gaol became an occupational hazard. In the 1840s he was a wholesale druggist and chemical manufacturer. By the 1850s he was in the iron trade as an iron merchant, but also

Albion Ironworks as seen from the towpath of the Walsall Canal, *c.*1860. (Source: Sandwell Archives)

Above: The interior of the Albion Ironworks, *c.*1860. The buildings are open at the side and fitted with slate roofs. (Source: Sandwell Archives)

Left: An illustration of the Britannia Ironworks, taken from Griffith's *Guide to the Iron Trade.*

became an agent for the London Bank and by 1861 he was publishing an iron trade circular from his offices in Darlington Street, Wolverhampton. He also used his skill with words to stand as a Liberal candidate for Wolverhampton in the 1861 elections for a Member of Parliament. He came second in the poll and was clearly popular amongst the working classes. As an ironmaster he worked the Bilston Brook Furnaces and Windmill End Furnaces, taking over the former from George and Alfred Hickman and the latter from George Pell. He also worked the three furnaces at Heyford, Northamptonshire, and the Britannia and Staffordshire Ironworks, West Bromwich. With the Staffordshire Ironworks, Samuel Griffith traded as E.B. Thorneycroft & Co.; he had persuaded Edward Bagnall Thorneycroft to let his name be used for the firm. In a vague agreement document, Thorneycroft was given the post of manager. Griffith was then able to capitalise on the famous Thorneycroft name.

Samuel Griffith was considered somewhat of a character and was associated with some colourful incidents. On 6 September 1860, he was involved in a fight with Henry Sparrow, in Birmingham, and both were injured. Griffith complained and Sparrow had to endure prosecution for assault in the magistrates' court. By 1862 Griffiths had already survived bankruptcy, in 1857, and with Thorneycroft faced another. A skilful Griffith tried to obstruct the prosecution by claiming the partnership between him and Thorneycroft did not exist. The court finally proved otherwise and the proceedings went ahead. Creditors pressed Griffith for payment in various court actions during 1862 and evaded the bailiffs through charm, and perhaps guile. He was arrested in April 1862 when it was believed he would leave Britain for Italy. Griffith was then marched around Birmingham to collect moneys to pay off debts. Later Samuel was laid up, through a dislocated ankle, either genuinely or by design, and avoided a court case in Stafford. These proceedings caused him to clash with solicitor, William Duignan, who acted for James Bayley for a debt owed. Shortly afterwards an untrue notice appeared in the local papers stating that Griffith was again in custody. The publicity had an adverse effect on his business and he sued Duignan. The resulting action was held in August 1862. Despite much of his former business career being made available to the court and public through the extensively reported proceedings, Griffith succeeded in getting damages from Duignan when it was proved that he had been responsible for the rumour.

Samuel Griffiths and E.B. Thorneycroft faced the bankruptcy court at the end of 1862 in proceedings that seemingly put an end to Samuel's activities in the local trade. However the irrepressible Griffith bounced back. His supporters saw him as a mediator in the subsequent puddlers' and miners' disputes and he was clearly a champion of the workers where his family roots lay. He subsequently moved to Winchelsea, where he was made a Baron of the Cinque Ports, and then London, where he set up another iron merchant's business in Cannon Street and edited a publication called the *London Iron Trade Exchange*. His exit from this world was just as original as the life he had lived. He died aged sixty-seven, on 24 May 1881, at Dalston Junction Station, North London Railway.

Fletcher, Solly & Urwick survived the drastic iron trade depression and Howard Fletcher retained the largest holding. When he died on 11 June 1861, the surviving partners were presented with a problem. His will dictated that his share in the operation should be paid to his children. Had they requested payment, serious financial problems could have arisen because the capital was tied up in the mines and plant. Henry Howard Fletcher and Benjamin Urwick were appointed executors of the Howard Fletcher's estate. They took charge of Howard's affairs and were responsible forming an agreement, in 1863, whereby the children retained Howard's share, but allowed the business to continue. Henry Howard Fletcher was the eldest son. He was educated Bridgnorth School and Cambridge University (BA, 1858) and, like his father, carried on business as a wine merchant. Henry also became interested in the affairs of the furnaces, principally through his duties as executor.

New Innovations in the Trade

During 1859, South Staffordshire blast furnaces produced 473,300 tons of pig iron. But even this relatively high figure was outstripped by groups in other parts of the country. In Northumberland, Durham and Cleveland the make was 617,966 tons. In South Wales and Monmouthshire, 985,290 tons were made and the figure for Scottish furnaces was 960,550 tons.

The greatest number of Black Country blast furnaces existed during the years 1860 and 1861. There were then sixty-six locations and 191 furnaces, but these figures included both working and closed-down furnaces. Many of the closed furnaces would never be restarted and were demolished. By then end of 1861 the number of blast furnaces at work had declined, only 108 remained in use. Bigger and better furnaces were being built, so the production of iron did not immediately suffer. Greater economies were also made to save fuel and to recycle iron lost in tap and flue cinders.

Innovations and improvements were made to the design and shape of the blast furnace, and the plant around the furnace became more complex and cluttered. The blast furnace was now slimmer and taller than previous versions. There was also a tendency to encase the whole furnace in iron, although there still remained a number of the older iron-hooped furnaces in daily use. The tops of the furnaces were altered to retrieve the waste gases. The hot gases were used to heat boilers and furnaces, the consumption of coal was therefore reduced. Various forms of mechanical haulage were also employed to raise the minerals to the furnace mouth.

The toll of ironworks sales continued with the sale of the Barborsfield Furnaces and Ettingshall Ironworks that belonged to Thomas Banks & Son. The plant then included two blast furnaces, a hot-air furnace, an incline plane to furnace and a condensing blast engine. The collieries extended to 48 acres, where coal and Gubbin Ironstone were mined.

Barborsfield Furnaces and collieries were part of an estate located on the north bank of the old Birmingham Canal between Capponfield Ironworks and William Baldwin's Bovereaux Furnaces. There was a large rectangular canal basin that had wharves for five boats. In 1861 the plant included both hot and cold blast pipes and hot-air ovens, indicating that one of the furnaces had been converted for hot blast. There was a casting house, instead of an open pig yard. The dimensions of this building were 50ft x 35ft, and it was equipped with a foundry crane. There was an incline to the top of the furnaces worked by an 8HP steam engine. A 70HP condensing steam engine with 40in-diameter cylinder provided the blast and three balloon boilers made the steam. The incline engine was also used to draw the pigs from the casting house to the wharf. Bank's fleet of canal boats was offered for sale at the same time and included a pair of wood cabin boats, a pair of iron cabin boats for the long-distance canal trade and several iron boats lacking the cabin that would have been used on short distance trips.

The separate Ettingshall Ironworks comprised mills and forges that made rods, strips, hoops, sheets and plates. There were six mills and fifteen puddling furnaces. The whole plant occupied some 3 acres of land. Following the sale, the furnaces were taken over by the Barborfield Co.

Edward Lowe Cresswell and his brother, Josiah Cresswell, were more fortunate and survived a call on their resources by creditors in 1859. They traded as Edward Creswell & Sons at Tipton Furnaces and Ironworks and were also partners in the Birmingham Cut Nail Co. In June 1859 they handed over their share in the Cut Nail Works to Alfred Neve Cresswell and concentrated on the iron business at Tipton.

John Jones & Sons operated furnaces at Green Lane, Walsall, and the 'Buffery' Furnace near Windmill End. This branch of Jones's family had long been associated with ironmaking in the Dudley and Brierley Hill district. There was no direct connection with George and John Jones of Bilston and Coseley Furnaces. At Green Lane there were two circular brick-built blast furnaces which were respectively 51ft and 52ft in height and each 14ft wide at the boshes. Compared to other local furnaces, Green Lane's were bigger than average and consequently had a greater make of iron. They were built of red bricks, lined with white firebrick and bound with wrought-iron straps or rings. Each furnace had six tuyeres and the raw material was taken to the top of the furnace by a pneumatic lift.

The two Green Lane furnaces were built for Thomas and Charles Highway in 1851 and 1853, to draw on the local supplies of Bloxwich coal and ironstone. Thomas and Charles Highway were already established coalmasters. Kelly's 1845 *Directory* lists them as proprietors of the Green Lane Colliery. Charles Highway was also associated with a flour mill in Wolverhampton Street, Walsall. The Highways fell victim to the failing iron trade and faced the problems that bankruptcy brought. For Charles Highway the pressure proved too much, one night, in 1860, when his family had gone to bed, he crossed the yard from his home to the flour mill, climbed the stairs and hung himself from a beam. The next morning the miller found his body hanging there. Few ironmasters were tempted, as Charles Highway was, to suicide, but bankruptcy remained at this time a great destroyer of lives.

John Jones & Sons took over Green Lane Furnaces in 1866 and traded there under the name of the Walsall Iron Co. The firm was actually a partnership of the sons of John Jones, that is: George Jones, Thomas Jones, Edward Griffin Jones, John Kenyon Jones and James Griffin Jones. The mines on the furnace estate had already been depleted. Raw materials were purchased from other nearby mines and frequently delivered to Green Lane by canal boat. Green Lane Furnaces were located opposite Birchills Junction on the Wyrley & Essington Canal. There was wharf accommodation on two sides of the works and a canal basin was made in the furnace yard.

Near to the furnaces was a large coke hearth that converted local coal into coke. These hearths were common to all Black Country blast furnaces, but at Green Lane they collected the waste gases and condensed them to get paraffin oil and ammonical liquor. Bloxwich 5ft coal was used to make the coke. Twenty tons of coal (about a boat load) would produce about 5cwt of oil and 2 tons of liquor. The oil was sold at 8*d* per gallon for lubricating purposes and the liquor was sent to local chemical works. This method was patented by Edward Jones in 1859 (patent 2,158), while he was employed as a furnace manager at Russells Hall Furnaces.

The Buffery Furnace, which had been formerly worked by Joseph Haden and later his son, William, was taken by Jones in 1868. It was similar to Green Lane being 50ft tall and 15ft at the boshes. This furnace was served by a branch of the Dudley No.2 Canal and, like Green Lane, used local coal and ironstone. This furnace is not to be confused with either the Old Buffery or New Buffery Furnaces, which were located on the colliery north of the canal. Joseph Haden seems to have been the first owner. Jones & Sons decided to economise on their coke consumption at the Buffery by using the slack which was normally thrown away to make coke. At the Bullfield Coke Works the small slack was ground into a powder, mixed with bitumen and burnt in special ovens to produce large cokes. These cokes were conveyed a short distance along the waterway to the furnace. A local name for the Bullfield Coke Ovens was 'Boshboil'. They were placed close to the old route of Dudley No.2 Canal. In later years this piece of waterway was reduced to a long basin, known as the 'Boshboil Arm'.

Coke ovens were particularly uncommon in the district, but those at Boshboil seem to have been erected as part of an experiment conducted by Edward Jones, ironworks manager, as an improvement in making coke in ovens. The method formed the basis for the patent presented in December 1858 (2,738), which received important publicity in *The Engineer* in 1859.

There was a tendency at this time to rebuild and replace old-style brick furnaces with iron-cased furnaces, erected with a growing concern for fuel economy. Tipton Green Furnaces were completely rebuilt between 1856 and 1860. Benjamin Gibbons junior and William Roberts had managed the works since 1851. The demand for pig iron had constantly increased and they decided to invest their money in new plant. Four larger furnaces replaced the original three. Built to the modern designs of the time, the improvements included a stationary steam engine to haul the barrows up to the charging floor.

William Ward & Sons also found a hitherto unexploited area south of Willenhall and erected the New Priestfield Furnaces. These works were established beside the Walsall Canal and were

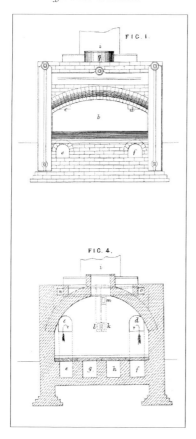

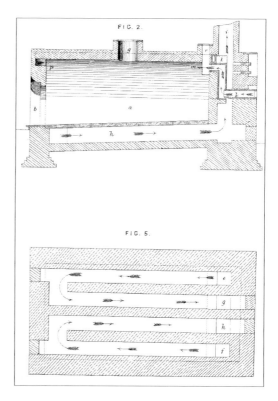

Above: Edward Jones's patent drawings,
2,738, for coke ovens, showing the
section across the furnace.

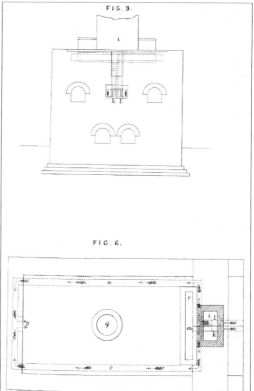

Above right and right: Edward Jones's patent
drawings, 2,738, for coke ovens.

nearing completion at the start of 1859. Two of the Ward sons had died young, leaving George (1833–1899) and Henry (1828–1904), and it was them that carried the company forward following the death of their father. Henry was the most active of the siblings; he married Jane Bagnall, the daughter of John Bagnall, ironmaster of West Bromwich, in 1855; was both a Justice of the Peace and Deputy Lieutenant for Staffordshire and was also a director of the London & North Western Railway. Henry's sister, Mary, married John Nock Bagnall, son of Thomas Bagnall, brother of John. New Priestfield came to replace the older Priestfield Furnaces and mines that eventually closed, drawing on the untapped mineral estate around New Priestfield.

For the Dawes family of Bromford Ironworks the decade provided new opportunities. John Dawes & Sons' ownership and management of this business had devolved down to the sons, John Samuel Dawes (1802–1878), William Henry Dawes (1805–1878) and George Dawes (1818–1888). Each brother pursued different careers, with John Samuel Dawes taking charge of the Oldbury Furnaces, formerly George Parker, during the early 1830s; William Henry Dawes operated the Withymoor Furnaces, near Netherton, and George Dawes took an interest in Yorkshire Furnaces.

John Samuel Dawes had the ability for invention and he patented a method for the use of hydrogen gas in blast furnaces, which was first tried successfully at the Oldbury Furnaces. John left the partnership at Bromford Ironworks in 1850, leaving his brother, William Henry, to continue the business. He also gave up furnace operation at Oldbury to Captain William Bennitt. John then pursued other ventures such as the difficult Manor Lane Colliery sinkings at Halesowen, and various patent inventions including agriculture and breech-loading firearms. He was also involved with brother, George Dawes, in mining in North Wales, near Mold.

Withymoor Furnaces became an important supplier of pig iron to Bromford Ironworks, which was despatched by canal. This traffic was aided through the opening of the Netherton Tunnel in 1858. William Henry Dawes was also the owner of Blackheath Colliery that supplied the ironworks and furnaces by canal. The Dawes family worked the Lyttleton Hall Colliery, developed on the lands belonging to John Dawes, the owner of Lyttleton Hall, which became the his home after he gave up the medical profession in Birmingham. Friars Park Colliery, on land leased from the Earl of Dartmouth, provided ironstone for the Dawes' blast furnaces.

The Dawes were also active in South Yorkshire, where the partnership of William Henry and George Dawes ran the Elsecar and Milton Ironworks that were served by the Dearne & Dove Canal and the Manchester, Sheffield & Lincolnshire Railway. A new venture in Lincolnshire was promoted during the early 1860s and completed in 1864, which depended principally on railway transport. The Trent Ironworks was established to exploit local supplies of ironstone. Three blast furnaces were constructed on land near to the Trent, Axeholme and Grimsby Railway and were fuelled by coal brought by the Manchester Sheffield & Lincolnshire Railway from the Duke of Newcastle's Colliery at Shireoaks, Nottinghamshire.

Blast furnaces for this period incorporated many improvements that led to increased iron production. Experiments were also being made to utilise the heat from waste gases, therefore economising on fuel. Lloyd's, Foster's and Willenhall's Furnaces were amongst the first to do so.

People were also keen to learn about their competitors' works and detailed descriptions started to appear in journals and books. A visit to the 'Quaker Works' in 1860 described Lloyd Foster's Old Park Furnaces at Wednesbury. This operation was fortunate in that local deposits of coal, ironstone and limestone were available. It is a graphic description of how iron was made:

> We cross to the three tall blazing furnaces. In their rear boys are breaking up the limestone into small lumps, under a range of sheds, and men are filling enormous wheel-barrows with roasted ore, with coal and coke, and the broken limestone. From the sheds a broad incline extends to the top of the furnaces, and up this the barrows are wheeled and placed one beside the other in a certain order, and the machinery being set to work, they are hauled up one side of the slope, while a train of empty barrows descends on the other.

We walk up and find our self on an iron platform, forty feet from the ground, and so near the blazing crests of the three furnaces as to induce a moments pause. The heat is however bearable, and we can walk around the craters as they may be called and look down upon the preparations in the foundry beneath, and listen to the roar with wonder. But the barrows are up; the 'filler' opens an iron door, wheels the foremost barrow across the platform, and shoots its load of half a ton into the furnace; then another and another, till all are emptied, when they are sent down to be refilled, an more ones are drawn up. The opening of the door makes us start back from the intolerable heat and glare, for there we look directly into the furnace, on a level with the summit of a huge mass of fire, forty feet in height, We can almost imagine it a volcano, so fierce is the heat, so angry the roar; and are impressed beyond previous conception by the tremendous forces required to make nature surrender her mineral treasures. And this goes on day and night, Sundays excepted, without intermission, for four years at a stretch till the hearth requires renewing, or circumstances necessitate a 'blow out'.

The ordinary charge is three barrows of coal, one of coke, two of ironstone, one of limestone, the latter being the flux, facilitating the separation of iron, as borax is the flux of assayers. The proportions vary according to the quality of iron required, and to maintain any particular quality, the order in which the barrows are shot must be carefully maintained. Charge after charge is shot in all through the twenty-four hours. Once in the furnace the fire lays hold of the crude mass, the fuel sinks and disappears, the iron yield to the intense heat and trickles down to the hearth, and accumulates there in readiness for the periodic discharge. Every twelve hours at five in the morning and five in the evening, tens tons of iron is drawn off from each furnace.

The puddling process continued to receive the attention of inventors in search of improvement. Various attempts took place to devise a steam-powered puddling method, while others concentrated on revising the design of furnaces. One original invention was devised by an Austrian named Jacob, who used hydrogen gas in the puddling process. For his experiment he used the Victoria Ironworks at Smethwick. Steam was passed through iron retorts which contained red-hot iron plates. The steam was broken down into the constituent hydrogen and oxygen gases. The hydrogen was collected in a gasometer and was sent from the puddling furnaces where it was burnt in a current of air. This method of heating was considered superior to that of using coal. However, the availability of local coal supplies meant that few ironmasters would consider adopting it.

Colliers Strike, 1864

Industrial actions became common following the major slump of 1857 as ironmasters and coalmasters reduced or maintained wages at levels which the workforce considered unjust. Coal was the most important commodity to the ironmaster, who required it for the smelting process, to heat the puddling and mill furnaces and to heat the boilers that raised steam for the engines that turned the mills, raised the minerals to the furnace mouth or provided the blast to raise the temperature in the furnaces. Miners withdrew their labour during the summer of 1864, seeking an advance on their rate. The dispute dragged on through to October, gathering pace. Meetings were regularly held and, as in previous times, some miners were determined to bully working miners to stay out. On one occasion a jar of gunpowder, with a lit fuse, was thrown into the home of Joseph Roberts at Lower Gornal. Roberts worked as a miner for the Earl of Dudley at the wage set, but the striking miners were keen to set an example. The gunpowder exploded, damaging walls and blowing off part of the roof. Fortunately the family escaped and Thomas Marsh was arrested and brought to trial for the act. John Hayward, who worked in the post of underground foreman, popularly known as a 'Doggy', for the Earl of Dudley at the Old Park Colliery, was similarly targeted. His home at Springsmire was attacked with a charge of gunpowder packed in a breakfast can. Some damage was done to a window and Francis Bennitt was eventually arrested for causing the explosion.

This, and many more acts, kept the local constables busy. The police force now had the responsibility of maintaining order. Their biggest problem was those bent on causing violence and disruption. The police monitored meetings and picked out the chief troublemakers. William Breakwell and Josiah Jones were identified as being amongst those who were particularly outspoken, a fact which was to aid their arrest after an attack on Mr Sheen, agent to the Himley 1-4 mines belonging to the Earl of Dudley. Breakwell and Jones were part of a mob that attempted to stop Sheen and other workers going to the mines. The mob had bludgeons and stones and threatened physical violence to stop the men. Sheen had a pistol and defended himself, threatening to shoot if the mob molested him and his group. Breakwell, Jones and others were arrested and sent to trial on charges of conspiracy and riot.

Many ironmasters and former ironmasters condemned the strike and the attempt by some activists to 'ruin' the ironmasters. S.H. Blackwell wrote a letter to the *Birmingham Post* acknowledging the attitude of the strike leaders and, in particular, Thomas Griffiths, for the Bilston Colliers. Blackwell noted that the men hoped that if the master was ruined, the works might pass into other hands. He urged caution as the action might lead to permanent closure. Samuel noted the success of the new ventures at Seend and Westbury in Wiltshire and Grosmont, near Whitby, where production costs were generally cheaper. As a result, good quality iron from Cleveland, Wiltshire and Yorkshire was reaching South Staffordshire mills and competing with the South Staffordshire brands. Iron was also arriving from Europe at cost effective rates. French ironmasters were delivering girders to London at £3 per ton less than South Staffordshire ironmasters. Belgian plates were also being delivered to London at rates cheaper than South Staffordshire quotations. The adverse affect on South Staffordshire ironmasters was clear; many blast furnaces were shut down. Those furnaces west of Dudley, for example, comprised only fourteen in work, out of a total of fifty-six.

Samuel Griffith made a point of addressing a miners' meeting at Bilston, pointing out that the struggle was hopeless while trade was in its existing condition. He specifically singled out the local pig iron trade, which was then in difficulties, and, in his opinion, in a worse condition that it had been for a long time. The ironworking trade, where pig iron was made up to merchant iron and plates, was in good condition. Samuel also noted that the railway companies had been active in procuring wagons to move coal from 'foreign' pits.

Some of the working miners also urged caution about the action of using foreign pits. Those in the trade were well aware of the limestone miners' fate, whose recent action had encouraged trade in limestone from quarries and mines further afield, such as at Froghall.

By the end of October the strike was effectively broken when the Coseley miners returned to work. Others followed, even though some miners stayed out. As with previous actions there was long term acrimony. Families were split and broken with father turned against son, or son against father.

CHAPTER 7

NEW HOPE IN THE TRADE, 1867–1880

Once the effects of the current iron trade depression and bank failures had diminished, the Midland iron trade headed for another upturn. This was a time of improving technology, with hot blast and tall iron-cased furnaces becoming standard. Improved ironworks design was organising methods of production, arranging puddling furnaces, annealing furnaces and shingling hammers and mills to the benefit of working practice – instead of the random arrangements encountered in early ironworks.

Blast furnaces developed during the 1850s, or later, adopted all the latest improvements. These included those around Willenhall and Moxley, which was perhaps the last part of the South Staffordshire region to be exploited. The district around Willenhall became a particularly active mining area after 1850. Willenhall is situated between Walsall and Wolverhampton and is best known for the lock trade. The manufacture of locks and keys provided employment for a large section of the population. Conditions of work were generally poor and accommodation for the workers and their families was, at best, basic. The government of the day undertook an extensive survey of working conditions, which appears in the *Parliamentary Paper Reports*. The inspector found a town of narrow courts and alleyways. Many houses inhabited by the working classes often had squalid living conditions. There were no underground drains and very few privies. Workshops often looked out on dust and dung heaps. Despite all the degradation, filth and squalor, the inspector found it remarkable that the women kept their houses clean. Even if the women wore rags, they tried their best to keep these clothes, and the clothes of their children, clean. Unfortunately, he could not say the same for the men and boys who worked; they were mostly dirty, independent of the smut of the forge.

The inspector makes no comment about the coalmines, but anyone who cares to browse the early census returns will discover that Willenhall people were also once employed in the coal and iron industries. Coal and ironstone were mined in this district throughout the nineteenth century. There were early mines at Coltham, near the Wyrley & Essington Canal. Later, mining was carried out throughout the parish, mostly concentrated in the areas bounded by Bilston and Darlaston. Established ironworks owners in those districts sunk mines in Willenhall parish when closer workings became exhausted. Most prized was the ironstone required to make iron.

Tramways often conveyed the minerals to the furnaces and several of these temporary railways crossed the parish boundary in Moseley Hole Lane and Noose Lane to reach the mines. Much of this expansion was carried out during the 1840s and 1850s when most of these tramways were laid down.

Three important firms operated in this area: The Chillington Iron Co., Wolverhampton; William Hanbury Sparrow (Osier Bed) and William Ward (Priestfield). All had tramways that crossed into Willenhall parish for the ironstone mines. With each passing year additional property was acquired and the rails extended nearer to the town.

The late 1850s and early 1860s mark the zenith of the Black Country iron industry, when the maximum number of smelting and puddling furnaces were at work. Yet this was also a time of financial turmoil in the iron trade with several local ironmasters facing bankruptcy. Perhaps the most notorious event of the period was the failure of the Birmingham Bank whose downfall was attributed chiefly to large loans made to the ironmaster Samuel Blackwell during the 1850s.

New Supplies of Ironstone

Staffordshire furnaces, in general, benefited from increased availability of ironstone and ore that was brought into the district by rail or canal, and the demand generated exploration for new sources. The brown ironstone from Northamptonshire was not discovered until the middle of the nineteenth century. In 1851, S.H. Blackwell conducted experiments with this ore and started to use it in his works at Russell Hall, and later the Bilston New Furnaces. Others followed his example.

George Pell was an early Northamptonshire ironstone quarry owner who, for a brief time in 1859 and 1860, was proprietor of the Windmill End Furnaces. It is perhaps significant that between 1859 and 1860 he had three boats registered on the Grand Junction Canal. George Pell was also a creditor of Edward Cresswell & Sons' Tipton Furnaces. Cresswell failed in 1859 and no doubt owed money to Pell for ironstone bought. Another creditor of Cresswell was the Heyford Iron Co. who remained an important ironstone carrier from Northamptonshire until the late 1880s. The Heyford traffic is mentioned in the 1885 Birmingham Canal distance tables. Special rates were afforded to this trade so as to make it competitive with the railway rates.

G. Bevan supplied ironstone from his mines at Blisworth. In February 1860 toll credit was granted to the Blisworth Ironstone Mining Co. on both the Warwick & Birmingham and Warwick & Napton Canals. By March 1869 Bevan & Co. had applied for a reduction of the tolls on iron ore to Birmingham.

From 1859 iron ore also began to pass from the Oxford Canal to Birmingham. Quarries had been established at East Adderbury, near Kings Sutton. They were worked by the Adderbury Ironstone Co. and at first sent all their ironstone by canal. The stone was conveyed by tramways to the Oxford Canal and then by the Warwick & Napton and the Warwick & Birmingham Canal to reach the Birmingham canal network and the Black Country Furnaces.

An Ironworkers' Union is Formed

The 1860s were a testing time for industrial relations. Several references have already been made to disputes between workmen and masters concerning pay. Often, such disputes were ruthlessly crushed, but with each defeat came a greater resolve on the part of the workmen. The greatest problem the workers faced was that they could be replaced. If a miner was on strike there were others who might fill his place. The miners' cause was also often undermined by the butty system, where several butties might be employed by an ironmaster, working in different pits and extracting ironstone or coal. Each butty used the truck system to keep his workers in line. Conditions were further undermined as there was also a turnover of the butties employed when the ironmaster dispensed with one team to replace it with another. Miners therefore had little job security and lived their lives accordingly.

For the ironworkers there was also a turnover of staff, but in one trade, that of the puddler, the men had a lever with the managers. A puddler's job was skilled, they had the knowledge to determine when the puddling process was complete and the iron was made malleable. If they went on strike then the ironworks shut down.

Masters were well aware of this state of affairs and did their best to keep the puddlers under control. If they left work early without completing their required number of heats, masters did prosecute them in the magistrates' courts where they faced fines. Pay and quality of iron were common causes of dissent. In the latter case, some iron was better to work than others and the extra labour and time needed led to many grumbles in the workplace. Iron made by the hot blast process was a particularly contentious issue as puddlers found this type of iron more difficult to work.

Localised disputes were contained, but in 1863 puddlers across Staffordshire and elsewhere came together in an organised action for better pay and conditions. Unfortunately, this was also a time of slump as iron demand had been affected by the American Civil War. Many masters responded by refusing to accept their terms. Local ironworks were shut down and considerable friction occurred between those in work and those who were not.

Works were stopped at Wednesbury Oak, belonging to Philip Williams & Son, after an action in which twenty-one puddlers and four ball furnacemen left their employment. Phillip Williams was chairman of the South Staffordshire Iron Trade and his firm took swift action to quell any further action. The men were taken before the Bilston magistrates and the puddlers were sent to prison for month. They argued that they had left work through hot blast iron being supplied to them – their employers, however, suspected that they had left after finding out that Phillip Williams was supplying bar iron to Messr Dawes, whose men were on strike. The four ball furnacemen were allowed back to work at Wednesbury Oak on paying court expenses, through the benevolence of Walter Williams, one of the partners. The puddlers endured prison confinement, but on release were treated as heroes by the local people and entertained to a good dinner.

The action progressed through the summer until August. The masters finally agreed to most of their terms in what was perhaps one of the earliest workers' victories. Their success provided a template for later workers' actions, such as the 1889 Birmingham Metallic Bedstead Workers' Strike, proving that organisation was the key to improving the lot of the worker.

Ironworkers across the country decided to form the National Association of Puddlers, Shinglers, Rollers and Millmen. Their first annual conference was held in Brierley Hill in May 1864, where seventy-eight delegates, representing 6,870 men, attended. Topics included the continuing issue of pay, but other items such as the advantages and disadvantages of mechanical puddling were discussed. Moreover, the union was a source of raising funds for workers in distress during strike periods.

The fledging workers' organisation faced a great test in 1865 when masters across the country opposed union demands. The resulting strike action became known as the 'Great Lock Out'. South Staffordshire men decided to end their action in April 1865, when W.O. Foster was chairman of the South Staffordshire Iron Trade – the North Staffordshire workers went on with their strike. The union continued with aid, such as providing the cost of fares for emigration to Australia and New Zealand.

The Joint Stock Company Act

The passing, in 1862, of the Act, which permitted joint stock companies to be formed, enabled some South Staffordshire ironmasters to set up as public companies.

The death of Samuel Mills had led to the Lloyd family taking over his extensive ironworks and furnace estate at Darlaston Green. In October 1864 the Darlaston Iron & Steel Co. was formed to run the iron and steelworks, blast furnaces, cement works and mines.

Dimmack and Marten, who traded as the Parkfield Co., attempted to follow their lead. An advertisement published in July 1865 was the first indication of their intention to do so. Directors for the new company included George Brocklebank, a director of the Patent Fuel Co. The new company was to be formed with a captital of £400,000 to take over the extensive ironworks and collieries at Parkfield, as well as certain forges and rolling mills contiguous to the Parkfields Estate.

This first attempt to become a public concern appears to have failed, but a second effort, in 1867, succeeded. Formed on 10 October 1867, the new Parkfield Iron Co. Ltd was created to work mines and beds of coal, iron and limestone at Wolverhampton and Sedgley and manufacture iron, steel and foundry castings, etc. The first directors were Caleb Bloomer, Edward B. Dimmack and Thomas S. Hatton. Henry Marten retained the post of general manager. The largest shareholder was Caleb Bloomer, iron merchant, from Hill Top, West Bromwich. He was the eldest son of Boaz Bloomer who, at the time, owned the Pelsall Coal and Ironworks.

A list of shareholders in 1868 indicates a wide range of roles within the industrial community, including:

George William Underhill, Wolverhampton, Iron Merchant
George Edwin Lambert, Bilston, Chain Manufacturer
Thomas Strange Hatton, Wednesbury, Miller
Edward Bindon Marten, Stourbridge, Civil Engineer
Thomas & Francis Allden Penn, Brewers
James Cadman, Bilston, Timber Merchant
John James Bowater, Hill Top, Oil Merchant
James Wells, Northampton, Coal Merchant
Noah Hingley & Sons, Netherton, Ironmasters
Joseph Hickman Pearson, West Bromwich, Iron Merchant
Stephen Thompson Sedgely, Ironmaster
Wright & North, Wolverhampton, Ironmasters
Charles Roberts, Pant nr Oswestry, Lime Merchant
Thomas & Samuel Siddaway, Quarry Bank, Shovel Manufacturer
Edward Holt, Brierley Hill, Coke Merchant
Simeon Bradley, Tividale Hall, Coal Master
Thomas Carmichael, Whitehaven, Iron Ore Merchant
William Owen Savin, Oswestry, Limestone Merchant
Thomas Bantock, Wolverhampton, Coal Master
Claridge & North, Bilston, Ironfounders
Thomas Hardy, Wolverhampton, Brewer

The limestone merchants from Oswestry and the Whitehaven ore merchant provide an indication of the distance and variety of sources from which the Parkfield Co. obtained its limestone and iron ore. The iron merchants and other ironmasters would have been Parkfield Co. customers, taking Parkfield iron for working up into finished iron products.

Other company formations included John Bagnall & Sons, the Chillington Iron Co. and the Pelsall Coal & Iron Co.

Railway Locomotives on Ironworks' Railways and Tramways

Local ironworks already benefited from the canal network that permeated the region and branched out. There were many branches and basins that reached to the centre of every works, enabling transport to the heart of every operation. Complementing the canals were the horse-drawn tramways and railways that served mines, bringing coal and ironstone to the furnaces, or the tramways that conveyed iron, limestone and slag around the ironworks.

The provision of private sidings that connected ironworks with the public railways greatly facilitated the movement of iron, coal and ironstone to supply the smelting furnaces, puddling furnaces and rolling mills. They increased the supply, bringing material from further afield to supplement dwindling local reserves. Railways were often preferred to cart and wagon haulage as heavy material such as coke, pig iron, rolled iron, ironstone and slag were moved in bulk.

Agenoria was the first steam locomotive to be used on a private railway in the Midlands. Built in 1829 at Stourbridge by the firm of Foster, Raistrick & Co., this engine was used on the level section of the Kingswinsford (or Shut End) Railway that linked the Shut End Ironworks and Furnaces with Ashwood Basin on the Staffordshire & Worcestershire Canal. *Agenoria* remained the only steam locomotive in this region for a number of years. It was not until 1846 that the Earl of Dudley employed a locomotive to move minerals from the Old Park Colliery to the Level Furnaces by a

One of the earliest locomotives to work on the narrow-gauge tramways in South Staffordshire was the engine '122', built by John Smith of Coven. The image formed the basis for an advertisement for the firm's products that was published in 1863. It is believed that this locomotive was built for the Willenhall Furnaces and was probably named *AJAX*. The photograph may have been taken in the Little London area where the company tramways crossed the road to reach nearby mines. (Courtesy Late S.H.P Higgins)

standard-gauge railway. This short line became the precursor of an extensive private railway system built up by the Earl's agent, Richard Smith, during the 1850s and 1860s. The Chillington Iron Co. was also believed to have employed a shunting engine in their sidings that joined with the Grand Junction Railway at Moseley, near Willenhall, where they operated blast furnaces from 1848. Other firms followed suit during the 1860s and 1870s, including Alfred Hickman, Lloyds Foster & Co., the New British Iron Co., Pelsall Iron Co. and George & Richard Thomas.

An important innovation was the use of steam locomotives on the company narrow-gauge tramways from 1862. Willenhall Furnaces and Chillington Iron Co. both used them and Parkfield was the third local firm to employ them. Two locomotives were known to have worked on the Parkfield tramways, particularly on the lines to the railway exchange sidings at Ettingshall and to the canal basin at Spring Vale. Their first engine was made at the Pot House Bridge Foundry, Bradley, by R. & I. A. Thompson, who had family connections with the owners of Parkfields. These locomotives ran on tracks of varying gauges, but generally the width of this gauge lay between 2ft 6in and 3ft 3in. Their use derived a considerable saving in the cost of horse haulage. Other tramway locomotives also came to be used by Addenbrookes at Roughay Furnaces, John Bagnall & Sons at their Bentley Colliery and the extensive railway system that belonged to the New British Iron Co. at Corngreaves.

Fresh Developments in the Local Iron Trade

A detailed account of the Parkfield Works was published in the *Mining Journal*, 25 December 1869. The five blast furnaces were described as 47ft high and between 11ft and 13ft diameter at the widest part, the bosh. All five were then in work, making an average of 110 tons of pig iron per week. Two furnaces were described as square; that is, four-sided but tapering upward. Two were round and hooped, the other was round and plated outside. Iron furnace design had changed considerably during the period of their operation and the plant at Parkfield reflects the transition. The original furnaces would have been square; later developments led to a preferred tapering, round design held together by iron-hooped bands. The most modern design for this period was the round furnace totally encased in iron plates.

The manufacture of iron used up vast quantities of coal. Several methods were already perfected that resulted in fuel economy. Waste gases were collected from the top of four of the Parkfield Furnaces to heat stoves for heating the blast and other purposes. Each blast furnace worked on the principal of forcing air through the burning mass of coke, ore and flux. The blast was created by a steam engine which pumped the air through the tuyeres at the base of the furnace. At first the air forced through was cold, but some benefits in iron production were achieved through warming the air beforehand; many furnaces had converted to the hot blast principal. The hot blast reduced the amount of impurities such as sulphur and phosphorus remaining in the iron. At Parkfield the air in the blast was heated to 800 degrees Fahrenheit, but in other furnaces, such as were then at work on Teesside, the temperature could rise to 1,600 degrees Fahrenheit.

Hot blast iron became the standard for the smelting of all iron. The concept, as developed by Neilson for the first hot blast stove using iron pipes, had been improved and adapted as a refractory lined, cylindrical structure used for providing the heated air. Pipes and stoves associated with the collection and reuse of waste gases may have added to the clutter around the actual furnaces, but the benefit derived from their use meant real money saving. They did not only heat the blast, but also heated the boilers that raised steam for the engines on site. The utilisation of waste gases was calculated to save about 500 tons of fuel a week. This equated to 26,000 tons a year. Part of the saving was 16,000 tons of small coals used to heat the boilers, while the other 10,000 tons represented fuel savings for the furnaces.

The material injected into the top of the furnace was a mixture of ore, coke and limestone flux. Coke was made, according the Staffordshire practice, through coal burnt in open heaps. Limestone from Dudley and North Wales was used as a flux, which was broken by a Blakes stone crusher. Native ironstone, foreign ironstone and cinder were mixed together with twelve parts raw coal and one part coke and the limestone. The mixture, loaded in barrows, was then drawn up an incline to the top of the furnaces, as the coal used was not suitable to heat the boilers.

In 1869 the mineral property was calculated to be around 240 acres. The bulk of this included the Old Parkfield Colliery (102 acres), New Parkfield (51 acres) and Ettingshall Lodge (31 acres). There were some twenty pits now in use on the property, but another ninety-four, the bulk, were disused. In 1868, 81,338 tons of coal was raised at the Parkfield mines. This equates to about 270 tons of coal a day. Another forty tons of ironstone was raised daily. Mineral extraction was handicapped through faulting of the mineral measures. The Lanesfield Fault divided the property on the east side, while several faults on the west side broke up the strata.

The leasehold Ettingshall Lodge Collieries that had been taken by the Wolverhampton and Staffordshire Bank in 1859 were, ten years later, being worked by the Parkfield Co. again. These mines lay to the north of the Lanesfield Fault. In 1869, levels were being driven through the fault to give access to 60 acres of Blue Flat Ironstone.

Some sixty years of constant mining had considerably depleted the mineral resources, yet sufficient minerals remained to keep parts of the mines in use for another ten or more years. Two were usually together as upcast and downcast. The *Mining Journal* article describes the operation at the Meadow Pits (Nos 17 and 18), which were then raising coal from the fireclay seam. No. 18 Pit was a downcast shaft and 90yds deep to the fireclay coal. The shaft was 7ft in diameter and lined with bricks. Miners working the coal commenced at the bottom of the seam, in what was called 'coal holers', and on the fireclay below it. They opened and worked a space 4ft high and often used gunpowder to bring down the mass above. Miners had to cope within the restricted height of the galleries. Their method of working was termed as a type of 'longwall advancing' – stone and rock from the seam were collected and formed into pillars to support the roof as they went along.

At the time of the visit by the *Journal* reporter, the working face extended 60yds to the southern boundary and 140yds northwards there from. Mines were constantly changing underground as new headings were driven to unworked areas. Roads rise and fall at angles as they followed the seams; few workings were actually level.

The typical means of carrying the coal out of the workings was by a small wooden tub with a single iron band to hold the coal in place. These tubs, which were locally called 'skips', were drawn along the pit railway by a horse or an engine. In the Meadow Pits these tubs were capable of holding about 8cwt of coal. An underground hauling engine was used at the Meadow Pits to draw a pair of skips to the bottom of the No.17 Shaft, which was the upcast. Generally, 50 tons of coal a day were raised from the No.17 Pit and 25 tons of coal brought up from the No.18 Shaft. The New Mine coal lay some 14yds above the fireclay coal. This comprised a 6ft seam and a 3ft 6in seam which were usually worked separately, with the upper being worked first. The New Mine on the Ettingshall Lodge Colliery Estate had been extracted many years before, but surviving ribs and pillars had just started to be worked again.

There were various engines on the surface, which were associated with the pits. The Sandy Gay pumping engine (Ettingshall Lodge) was a beam engine (58in-diameter cylinder) that had been used to so much effect in the 1850s but now was employed only during the wet season. The daily pumping duties were carried out by another engine erected nearby. This engine, which was put into use about 1861, was of a modern design and had a 40in horizontal cylinder with a 5ft stroke.

Five steam winding beam engines, or whimsies, and eight horse gins were used to raise the minerals. The Ettingshall Lodge Whimsey drew from two pits: the New Mine Pit and the Blue Flats Pit. The Lanesfield Whimsey also drew from two pits, one to the Thick Coal and one to the New Mine Coal. The No.1 Whimsey, near the ironworks, drew from four pits; two to the Gettin Rock, one for the Fireclay coal and one for 'repairing'. An atmospheric (Newcomen type) engine served two pits; one to the New Mine and one for the White Ironstone. The Meadows Whimsey served the Nos 17 and 18 Pits. There was also a hauling engine at the top of the No.17 Pit that worked endless chain haulage through two roads in the No.17 Mine.

Edward Dimmack's involvement with the Parkfield Co. had been diminished after his bankruptcy proceeding of 1859, yet he still kept close ties, but these lessened over the years. One of his last acts was made in 1870, when he handed over to Henry Marten, by deed of assignment, sundry assets including his share of sixteen canal boats.

Pig iron was the main product of the Parkfield Iron Co. Although there was a brief period working the ironworks at Caponfield, the bulk of their funds were provided by the sale of pig iron. This was a highly competitive business. Not only was Parkfield competing with its fellow Staffordshire ironmasters, but also the growing number of furnaces elsewhere. The South Staffordshire Ironworks were buying an increasing amount of 'foreign' iron, which came into the district by boat or railway wagon.

At the start of 1871, the South Staffordshire Ironworks were doing little business, but the work picked up from March. There was such a fine line between success and failure that slack times such as these could precipitate bankruptcy. Caleb Bloomer was then chairman of the Parkfield Co., but other commitments led him to resign in April 1871. Caleb retained his shares, but other shareholders left the company. This simple act was to precipitate a crisis.

The Parkfield Co. was then considered one of the most prestigious firms in the country, yet such was the nature of their business that when credit suffers, investors naturally try to safeguard their interest. An extraordinary general meeting was held on 28 July 1871, when it was decided to wind up the Parkfield Iron Co. Benjamin Smith of Wolverhampton was appointed liquidator.

Ironmaking during this period was complex. Each firm produced their own special brand of iron, which varied in quality. Often production was tapered to suit a particular market. The Parkfield Co. had chosen to enter the most competitive market, selling the cheapest form of iron.

When prices fell, the firm's credit suffered while other costs remained at the same, regulated prices. Parkfield was bringing ore from Cumberland to mix with the local ironstone. The cost of the ore and transport was such that the Parkfield Co. could scarcely cover the expense through iron sales.

It was announced at the meeting of creditors that liabilities stood at £9,000, while assets were £7,000. These figures were encouraging enough to carry on the firm, and Henry Marten was determined to continue trading. New capital was brought in. Caleb Bloomer increased his shareholding and other new investors included J.R. McClean and Matthew Frost. Mortgaging the Parkfield Works raised another part of the necessary capital; this agreement was made in March 1873 with the Wolverhampton and Staffordshire bank to secure £7,641 6s 5d. Both the names of Henry Marten and Edward Bindon Marten appear on the mortgage.

Iron production was carried on through 1873 and 1874 but eventually the Parkfield Co. had to admit defeat and allow liquidation proceedings to go forward. Parkfield became the property of the main creditors, the Wolverhampton and Staffordshire Bank, who instigated the sale of the works and mines. Parkfield Ironworks and Collieries were offered for sale by public auction on 9 February 1876. The sale included five blast furnaces with incline engines, hot-air ovens, gas apparatus and boilers. South Staffordshire Mines Drainage Commission Control had taken over the operation of the Sandy Gay pumping engines on the Ettingshall Lodge Colliery.

Nathaniel Neal Solly used his skill to guide the Willenhall Furnaces through these changing times. Each furnace was initially 45ft high and 13ft in diameter at the boshes. He and his partners used their profits wisely, investing in new plant where practical. From 1860 the furnace gases were used to heat the blast furnace boilers and from 1861 were also used to heat the furnace blast. Their manager, R. Griffith, made some important arrangements, such as the rebuilding of No.3 Furnace with a Lurmann closed hearth. The hearth was 6½ft wide and designed to concentrate the heat – it

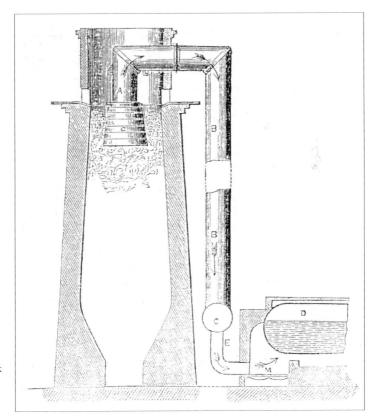

A diagram illustrating the collection of waste gases as adopted by Fletcher, Solly & Urwick at Willenhall Furnaces. (Source: *Mechanics* magazine)

led to an increased production of up to 210 tons of iron per week. The waste gases were taken off the furnace according to the principal devised at Roughay by Addenbrooke and Millward. No.1 Furnace retained the old style hearth and employed the Darby Bell method of taking off gases. Two kilns were employed to calcine the Staffordshire ironstone and this was raised to the top of the furnaces by an incline.

They used a mixture of two-thirds coal and one third coke with the ironstone and lime. The New Mine Coal is used and the Bottom Coal is coked. In 1870 five pits were employed to raised coal, and others for ironstone, Robins, Balls, Gubbins and Blue Flats were extracted.

In 1876 Solly and Urwick decided to arrange for a public company to work the concern. The new venture became known as 'Willenhall Furnaces Ltd'. Henry H. Fletcher, N.N. Solly, James Solly and Benjamin Urwick were appointed directors. The capital of the company was £50,000 and the directors set about making further improvements. The heights of the furnaces were raised by 10ft, furnaces were repaired, hearths rebuilt and a standard-gauge railway connection was made with the Walsall & Wolverhampton Railway. Minerals in the company mines were reaching exhaustion and the railway connection was seen as a means to continue supply of minerals. The siding was opened for traffic in October 1877.

The Addenbrooke brothers, George and John, guided the Rough Hay Furnaces from survival to success through careful management of their business. Improvements continued to be made to the furnaces and the brothers benefited through the capable attentions of their engineer, P.A. Millward. Both he and John Addenbrooke were responsible for a number of innovations at Rough Hay and at their mines and furnaces at Bedworth, Warwickshire. Three calcining kilns were built at Rough Hay to 'calcine' (or heat with coal) the ironstone before it went to the furnaces. These kilns replaced the earlier system that hitherto existed, where the ironstone was calcined in open mounds. Rough Hay was believed to be the first to adopt this method as kiln calcining was previously believed to be unsuitable for treating the local ores. Addenbrooke showed that it was possible and of benefit to ironmasters as there was a saving in fuel. Another invention was a method of collecting waste gases from the furnace mouths that was patented by the Addenbrooke brothers. They also adopted locomotive haulage on their colliery railways, with what is thought to be a 3ft 3in-gauge railway that brought coal and ironstone from the pits to the furnace bank.

The Bedworth Collieries were served by the Coventry Canal, which first linked Coventry with these mines and was gradually extended until this isolated waterway became part of the national network. The Addenbrookes became associated with the Bedworth Charity Mines and eventually formed the Bedworth Coal & Iron Co. Ltd. The estate, in 1873, included some six pits for drawing ironstone and another two for coal. During 1873 two blast furnaces, 50ft high, were erected on the colliery land close to the canal basin that joined the Coventry Canal. Private railways linked the mines, furnaces and canal. The blast furnaces were served by a lift that carried materials to the top.

Darlaston Iron & Steelworks experienced periods of mixed fortunes. They had blast furnaces beside the Walsall Canal, extensive ironworks with mills and forges, a steelworks with steel converters as well as coal and ironstone mines. Their property included the rich mineral property at Essington Wood, which Samuel Mills developed for canal transport to Darlaston and which the Darlaston Iron & Steel Co. improved with a railway link in 1875. This operation had been a partnership of Richard Bills and Samuel Mills until February 1849 and was then in the sole ownership of Samuel Mills until 1863. The public company promoted by the Lloyd family, in 1864, lasted until 1877 when losses in trading led to a break up of the concern. The Essington Wood collieries passed into the control of the Darlaston Coal & Iron Co. Ltd. A reconstituted Darlaston Iron & Steel Co. carried on the Darlaston operations until 1882, when bad trading again led to further disposals.

Failures continued to be a facet of the trade. At the start of the 1865 the chief topic of the local Iron Exchange was the suspension of William Edgeworth Gibbons, who had lately been carrying on the Great Bridge Ironworks, formerly in the possession of Samuel Griffiths, and who also

Engraving showing the single, and original, furnace at Darlaston Green as seen from the West Canal Basin. (Griffith's *Guide to the Iron Trade,* 1873)

Engraving showing the Darlaston Green pig yard (East Canal Basin) with Crowley & Co. boat. Sales plans that show the location of these furnaces are reproduced on page 157. (Griffith's *Guide to the Iron Trade,* 1873)

had been a maker of pig iron at Deepfields, where he had three blast furnaces in his possession. A meeting of creditors found liabilities in excess of £14,000, but assets scarcely more than £3,000.

New methods of working and new types of products heralded a time of change for the South Staffordshire ironmasters. The changing nature of the trade led to a panel of ironmasters being appointed to act as a committee to the iron trade in January 1865. These were William Mathews, John Hartley, William Barrows, George Barker, Walter Williams junior, Frederick Smith and William Hanbury Sparrow. Each of these gentlemen were to be chairman for a year, in rotation of their names.

The year 1867 marked another downturn in trade, at a time when British products were winning medals at the Paris Exhibition. Bradley sent a case full of products that won medals, but the honour was not enough to prevent his creditors forcing him into bankruptcy. George Beard was a partner in the Regents Ironworks and went on to form a new partnership with B.H. Eberhardt at Grove Ironworks, Cape and Eyre Street Ironworks and Springhill, Birmingham.

Concern for expenditure was leading more ironmasters into collecting furnace gases to heat boilers. S. Whitehouse of Kates Hill developed a device that involved the fitting of a cupola. They sold the device to William Fleeming Fryer, banker of Wolverhampton, who owned the Hatherton Furnaces. Whitehouse made and installed the apparatus and also patented the idea. The agreement for making and fitting was set at £130, which was duly paid. Then Whitehouse decided to charge another £100 for the patent rights. The matter went to court after Fryer sued for £130 plus damages when the cupola fell into the furnace. Despite the problems with Hatherton, S. Whitehouse went on to sell similar devices to H.O. Firmstone and Phillip Williams & Sons. A different type of collection device was fitted at one of the Rough Hay Furnaces. In 1867 George Addenbrooke, one of the proprietors, reported that after two years of operation the production of the furnace had increased to nearly 200 tons of pig iron per week.

Despite increasing numbers of failures in the South Staffordshire trade, the Round brothers decided to erect blast furnaces at the Hange Collieries during the year 1867. The Round family had hitherto been engaged in the coal and ironstone trade. John Round built up a coal and iron business based principally on good supplies of ironstone at the Hange. The name 'Round' was fairly common in this district and there were Rounds involved in both the coal and timber trades. This particular John Round lived in Daisey Bank, near Princess End, and married Phoebe Caddick, the daughter of Isaiah Caddick. He was a partner with Caddick and Zachariah Parkes in the Old Church Ironworks at Tipton and gained an increased shareholding in this ironworks when Parkes died in 1821. The estates at the Hange were leased principally from the Duke of Sutherland and Round paid a royalty on all the minerals raised and traded from the Hange as both coal and iron merchant.

Control of Hange collieries passed to Daniel George Round. Daniel, born 1820, was the eldest son of John Round. He was a magistrate for Staffordshire and Worcestershire and lived at the Hange and then Portland House, Edgbaston, near Birmingham. A massive Cornish Engine, known as the 'Neptune' engine, was supplied in 1851 by Hockin & Loam of the Perran Foundry, Truro, to drain the Hange Colliery. Barrows & Hall of the Bloomfield Ironworks supplied three boilers.

The Baldwin family had close connections with the Bilston area through their engineering works, Bovereaux blast furnaces, collieries and tinplate works. They owned mineral property that extended from the banks of the Old Birmingham Canal through to the Bilston Branch of the Walsall Canal. Proprietors of this estate were the Baldwin brothers, who included William and Martin. They were baptised at Madeley, Shropshire, as the sons of William Baldwin and Anne Pugh.

William Baldwin became one of the Bilston Township Commissioners who created the Bilston Improvement Act of 1850. William, who lived at the Ellowes Hall at Sedgley, along with fellow coal and ironmaster, Edward Bagnall Dimmack, helped preside over the important changes needed to improve services to the town. The commissioners tackled such issues as water supply, sewers, housing standards and the provision of a cemetery. Nine acres of land for a cemetery were purchased

The Cape Arm of the BCN provided wharfage for several ironworks. The Regents Grove Works, shown above, fronted the top part of Grove Street. (Griffith's *Guide to the Iron Trade*, 1873)

from William Tomkinson Riley, ironmaster. They also secured an improved water supply through purchasing water from certain ironmasters, such as John Bagnall of the Capponfield Ironworks. In 1853 William Baldwin agreed to allow the public use of his private road through Bovereaux Colliery to Glasshouse Bridge on the Birmingham Canal.

William Baldwin died about a year later. His will, dated 25 May 1854, provided some £8,000 to be used in the building of a new church at Bradley. Martin Baldwin continued the business and was the last survivor of the family business. The *Bilston Herald* recorded his death in February 1872, aged eighty-three, and mentioned that he and his brothers were architects of their own fortunes. Bovereaux Furnaces were subsequently taken over by Thomas Holcroft who traded under the name of Tame Iron Co.

George Haden Hickman and Alfred Hickman had acquired the Stonefield Ironworks during 1857 and had mortgaged them to Daniel Pritchard and Richard Palin of Shrewsbury. Following their receivership in 1861 the property was split, with the furnaces passing to the Stonefield Iron Co. and the ironworks to Joseph Sankey of Bilston, ironmaster; John Page of Bilston, manager of ironworks and Richard Chambers of Bilston, ironmaster, as the leasees. They acquired the engine, plate and sheet mills, puddling furnaces, forge trains, helve hammers, squeezers, sheers machinery, building and fixtures. The lease contained a detailed description of the plant, which is reproduced below:

Details Stafford Record Office D643

First Schedule Stonefield Ironworks

   Condensing beam engine on cast iron frame cyl 38 in dia, stroke 6ft, cold water pump and pipes up to well two feed pumps and mains of pipes up to four boilers steam pipes up to same governor and throttle valve all complete up to end of spear rod including the wood and iron foundations – two egg end boilers and gearing complete with brick seating – cylindrical furnace boiler and gearing complete on cast iron standards – cylindrical firing boiler in brick seating with gearing complete fire doors bars bearers and buckstairs – driving gearing complete consisting of main shaft nineteen feet long by fifteen

Regents Ironworks on the old Birmingham Canal at Bradley belonged to a group of various forges and rolling mills that lined the canal in this district. The open day boat seen in this view was the common form of transport that was used for moving coal, ironstone, pig iron and finished iron products.

Eyre Street Mills was one of the few works located within Birmingham that had puddling furnaces for working-up pig iron into the malleable form. (Griffith's *Guide to the Iron Trade*, 1873)

Bilston Iron Co., Stonefield Ironworks. Stonefield comprised a blast furnace, puddling furnaces and mills. These works were placed at a distance from the Birmingham Canal but had road access to a wharf near Capponfield. (Griffith's *Guide to the Iron Trade*, 1873)

inches square having thereon a sixteen feet driving wheel large crank governor pulley and chain, fly shaft ten feet long by seven inches square having thereon two pinion wheels and sixteen feet fly wheel – shaft for working little mill six feet six inches long by twelve inches square having thereon a chain pulley a crab a seven feet driving wheel and a ten feet driving wheel. Shaft for driving boiler plate shears seven feet long by eleven inches square having thereon an eccentric cam and a driving wheel eleven inches in diameter – shaft for cutting down shears six feet six inches long by eleven inches square having thereon a circular eccentric cam – eleven feet driving wheel crank and crab – four long cast iron beds on top of silling with new large carriages with glands holding down pins – strong and massive wood framing and foundation under the whole of the above driving gear – wrought iron water tank pipes and cast iron columns – pair of vertical squeezers bed plate and foundations with wood iron bound connecting rod complete – forge consisting of helve two standards and bed plate cup block anvil hammer cam ring with three arms standard for same crab and strong timber and other foundations complete – an eighteen inch forge train complete consisting of spindles, couplers and crabs pair of pinion housing with pair of pinions fitted with chocks etc five spindles and nine couplers two pairs of housings with screws and boxes and three spanners – two pairs of eighteen-inch rolls for ten-inch bars and six-inch bars – long cast iron bed plate – three wrought iron cramps and various wrought iron pins. Set of chocks to each housing – wrought iron standards for roll levers – wrought iron water trough two fire plates – an eighteen inch sheet iron train complete consisting of spindles couplers and crabs – pair of housings with screw boxes chocks and two spanners – pair of chill rolls – barrel four feet long – pair of housings with screws boxes chocks and two spanners. Pairs of grain rolls three feet eight inches barrel fur spindles and seven couplers pair of pinions with housing chocks wrought iron pins etc – long cast iron bed plate – two cat iron fire plates – frame standard cramps and lever bars – powerful boiler plate shears with arm and strong fixings complete – pair of vertical cropping shears and wrought iron connecting rod complete – pair of vertical cutting down shears and wrought iron connecting rod complete – shearing machine for nail sheets with fly wheel band pulley and overhead gear complete – large cutting down shears with long horizontal arm bed plate holding down pins and foundations complete – a twenty inch boiler plate train complete consisting of spindles couplers and crabs – pair of housings with seven boxes two spanners glands and cramps complete – pair of chilled rolls barrel four feet long the top roll being twenty two inches in diameter – six spindles eleven couplers and two crabs – pair of housings with screw boxes two spanners glands and cramps complete-seven chocks to the four housings – cast iron frame and levers for top grain roll with wrought ironwork complete. Pair of grain rolls barrel four feet two inches – pair of pinions with housings chocks, glands and wrought iron pins complete – pair of bar rolls barrel three feet six inches long with pair of housings with screw boxes ten chocks etc cramps and glands complete – long cast iron bed under the train three fire plates – four puddling furnaces with stacks complete – eight ditto ditto without stacks working into egg boilers with flues and brickwork complete – one ditto ditto working into large engine stack complete – small mill heating furnace and stack complete ditto annealing ditto – large mill heating furnace with stack complete – ditto, ditto working into cylindrical boiler – ditto annealing furnace with stack complete.

Buildings blacksmiths shop and hearth with iron plate thereon complete – brick built storeroom and room over same – pig iron yard enclosed with walls and iron palisades, scrap yard enclosed with walls, brick built warehouse with stated roof – brick built stable with iron roof (now occupied by the Stonefield Iron Co.), brick built carpenters shop, brick built clayhouse and old carpenters shop, iron watch box, wood and sheet iron roof over three puddling furnaces on eight cast iron columns wood and slate roof over three puddling furnaces on seven iron and two wood column – wood and sheet iron roof over one puddling furnace on three iron column – wood and sheet iron roof over nine cast iron and two wood columns over four furnaces – wood and slate roof over one furnace on one iron column – wood and slate roof over three on four cast iron columns – large wood and slate roof over forge and smaller mills with ventilating ridge with six wood principals on eight iron and four wood columns – single slant wood and slate roof over the shears on three round cast iron columns – large gable roof (wood and slate) over driving gear and large mill with four principals, two iron, one wood and one brick column – small tiled roof on for wood and one iron column at front of heating furnace – iron hovel lofty brick built stack to firing boiler, ditto ditto two egg end boilers – brick built tipping wharf and pens for red ore with iron doors and frames – cast iron floor plates about three thousand six hundred square feet, wrought iron floor plates a quarter of an inch thick about one hundred and ninety square feet. Wrought iron water pipes with taps and fittings running through worked to the trains, machinery necks and to the forge and mill furnace complete. Weighing machine to be removed from the Queens Garden Colliery complete.

Second Schedule Stonefield Ironworks

Blocking and fixing all wheels throughout the demised works and premises – putting down trains of rolls throughout said works and premises putting roofs in thorough repair – erecting and completing a proper fence wall of brick or stone – repairing engine – repairing all the puddling and mill furnaces – supplying brasses throughout the works where needful – repairng boilers – repairing furnaces – removing and re-erecting mill furnace – removing from the queens garden colliery and refixing weighing machine – securing holding down pins.

Joseph Sankey was better known for making holloware and developed an extensive business around the town.

Summerhill Ironworks were owned by family firm, the Millingtons. They had taken over parts of the Tibbington Estate and continued first as Thomas Millington, then as Thomas Millington & Sons, William and Isaiah Millington and finally as William Millington & Co. The establishment came to include sixteen puddling furnaces and rolling mills.

William Bennitt, magistrate and captain in the Dudley Squadron of the Queen's Own, became involved with both the coal and iron trades and by his marriage to Sarah, daughter of ironmaster John Dawes, developed close ties with the Dawes family. He took over the management of the Oldbury Furnaces, but in 1866 was bankrupt and the Oldbury Estate was offered for sale. The furnaces passed to the Onions family, while William Bennitt's son, Henry Pycroft Bennitt, continued to work the adjacent brickyard.

The name 'Onions' was once a common name in the iron industry, although no obvious connection can be traced between them. There was the Onions family of Broseley that were associated with the Bancks in iron manufacture; the Onions of Kilsall, which William Firmstone married into, and the Onions of Birmingham who established a bellows trade and later, with the Alldays, engaged in general engineering and the automotive trades. There were also the Onions associated with the blast furnaces in South Staffordshire from the 1860s. Their tenancy

Engraving that shows the Summerhill Ironworks, belonging to Millington & Co. at Tibbington. (Griffith's *Guide to the Iron Trade*, 1873)

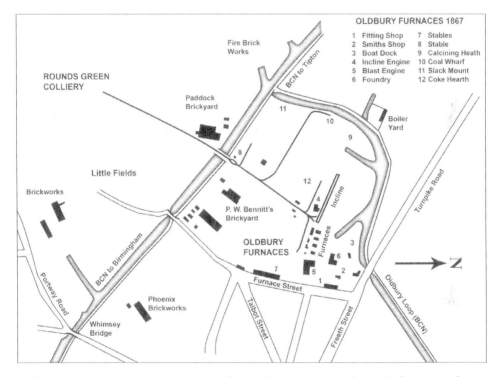

Oldbury Furnaces, *c.*1867, showing the tramway, foundry, furnaces and the incline to the furnace mouth.

was somewhat nomadic, moving from furnace to furnace, treading a fine line between solvency and insolvency. John and William Onions were partners in the Windmill End Furnaces until May 1865. Edwin, Joseph and William Onions had charge of Bradeshall Furnaces. Joseph left this firm in April 1867.

Brades Hall Furnaces, also some times known as 'Stour Valley', had been established by Benjamin Richards on a strip of land adjacent to the Birmingham Canal New Main Line and the Gower Branch in 1853. It was an extremely isolated position, suited to canal transport, but difficult to reach by road. The bulk, if not all, of the raw materials were brought by boat to be worked up into pig iron. Richards had filed, and survived, a petition for insolvency during the crisis of 1857, but failed again in 1860 and subsequently one of the two Brades Hall Furnaces passed to the Onions. Another ironmaster, Frederick Giles (of the Portfield Ironworks, Dudley Port), briefly worked the other Brades Hall Furnace until financial difficulties led him to give it up. The Onions then operated both furnaces.

William and Edwin Onions had charge of the Oldbury Furnaces until 1872 when creditors pressed for payment and the Onions gave them up. Oldbury Furnaces comprised four blast furnaces and were capable of making 600 tons of pig iron. Associated with these works were coke and calcining hearths, an incline, fitting shops and a foundry. The sale of the works in August 1872 evidently found no prospective buyer and materials of the furnaces were put up for sale in November 1872, including the brickwork, tuyeres and the wrought-iron bands for the furnaces. These furnaces were dismantled and the Great Western Railway subsequently adapted the land as a canal and railway interchanged basin.

James and George Onions also tenanted Brades Hall Furnaces for a time before they established a brand new furnace and ironworks alongside the Dixons Branch, Tipton, in 1870, that became known as 'Horseley Heath'. This new Onions' furnace belonged to the new generation of furnaces where the waste gases were collected to heat the boilers. An 110ft-long incline plane conveyed materials to the top of the furnaces and a large basin in the furnace yard had accommodation for between seventeen to twenty boats, reinforcing the continued requirement for canal transport. James and George Onions of Keeling Street, Tipton, went before the bankruptcy court in 1871 and their furnace was put out for a time. It was restarted under the proprietorship of the Dudley Port Furnace Co.

More detailed descriptions of the Black Country furnaces, published in the *Mining Journal* in 1870 and 1871, show that several, if not most works then used a mixture of local and imported ores. The heamatite ores were favoured for their high iron content. The New British Iron Co. used the red heamatite ores, the brown North Staffordshire ores and the local argilaceous ironstone in their six blast furnaces at Corngreaves. G.B. Thorneycroft smelted a complicated mixture of Northamptonshire, North Staffordshire and north of England ores, in addition to local ironstones, at his Bradley Furnaces. Corngreaves had railway sidings at the time and the ores were no doubt delivered by rail. Thorneycroft's furnaces, however, had no rail connection and all ore must have been transported by canal for at least part of their journey.

Oxfordshire ironstone was important traffic and in 1870 ironmasters were able to negotiate a reduction of tolls on the Warwick & Birmingham Canal. They agreed to send 200 tons a week during spring and summer months (1 March to 1 September) and 150 tons a week during the autumn and winter months (1 September to 1 March). A similar arrangement was agreed for the Warwick & Napton Canal, which totalled 9,100 tons from 1 March 1870 to 28 February 1871. It seems that this target was not reached for the year 1873, only 7,876 tons had been sent, up to 28 February 1874.

Boaz Bloomer & Sons was converted into a limited liability company at the end of March 1873, at a time when the business was in profit. The new name chosen was the 'Pelsall Coal and Iron Company'. Their mines were producing 100,000 tons of coal per year and 15,000 tons of pig iron. The bulk of this was used to produce some 17,000 tons of wrought iron, the remainder of the coal was sold on. They had built up an extensive mineral property surrounding the ironworks of two blast furnaces, forty puddling furnaces six rolling mills, mines at Pelsall and Short Heath. A capital of £240,000 in turn generated additional capital for the purchase of further mines from the Charles family, this enabled the sinking of the No.10 Pit in 1874, and later estates from the Fishley Charity enabled the sinking of the No.12 Pit in 1885.

Directors of the Pelsall Coal and Iron Co. in 1873 were Boaz Bloomer Jr, George Hastings, Stephen Stokes, Dr John Watts and James Bissell. Boaz Bloomer junior, who had been employed as manager of the firm, became chairman of the board. His father, Boaz senior, retired from the business and died in 1874.

Canal boats and railway wagons brought ore and ironstone from other areas, in addition to mining coal and smelting local ores. Pig iron, too, was transported from a variety of locations, including the Brymbo Co., Lilleshall Co., Lincolnshire Smelting Co., Newton Chambers, Sheepbridge Coal & Iron Co. and the Wigan Coal & Iron Co.

Boaz Bloomer's other sons fashioned out careers in the iron and coal trade. Caleb Bloomer was in business as an iron merchant and a chain and anchor maker at Great Bridge and West Bromwich for over twenty years. But his other ventures proved less successful. He purchased an ironworks at Wolverhampton that belonged to Mr Rose and was also associated with the ill-fated union Iron & Coal Co. The speculation led to his bankruptcy in 1872. George Bailey Bloomer chose the coal trade, here he was associated with his father-in-law, Mathew Frost, in mines at Bilston and Moreton Hall Colliery, near Chirk.

Smethwick was noted for ironworks and engineering, but this area also briefly had a blast furnace at the Cape Works, owned by B. Richards. During October 1869 the stock of calcined cinder and Northampton ore was for sale at the Cape Furnace.

During September 1873, the Solly Brothers' Leabrook Ironworks were offered for sale, including a 16in forge train and ten puddling furnaces. These works were alongside the Walsall Canal and bounded by John Bagnall's Leabrook Works.

The Lees family, Samuel, and sons Samuel and John, moved from Tipton to Hill Top, West Bromwich. The two brothers followed completely different occupations. Samuel Lees junior became a shopkeeper and later a butcher, while John Lees became a butty collier. The Lees were Weslyan Methodists and John Lees is credited with building the church at Harvills Hawthorn. John Lees added the timber trade to his business ventures. Lees eventually set up a timber yard at Swan Meadow, near Swan Village. He then went into partnership with Edward Holden at the Brickhouse Ironworks that made gasometer and boiler-plates, pan tray iron as well as the general types of rolled iron, bars and strip.

John Lees died on 25 May 1869, leaving a widow, three sons and two daughters. The sons, John Bayley, Samuel and Charles took over the management of the Swan Meadow Timber Yard and their father's share in the Brickhouse Ironworks. The ironworks was given up in preference for a new venture at the Albion, West Bromwich, which had been part of the works carried on by Walter Williams until 1860.

Noah Hingley was a pioneer in the manufacture of chains, cables and anchors. He began making chains in 1820 and established a large foundry and ironworks at Netherton in around 1845. These works were located beside the Dudley No.2 Canal and, from 1848, the anchors were made there. Hingley expanded his business and began to manufacture his own pig iron and mine coal. Blast furnaces were acquired at Dudley Wood and Netherton, which had formerly been worked by the New British Iron Co. Both plants comprised the old square furnaces. In 1871 the Old Hill Furnaces were taken over from David Rose and these became their sole iron-smelting plant. Dudley Wood closed down and Netherton was sold.

Thomas & Isaac Badger of Dudley made their fortune through the nail ironmonger business. They acquired land at Rowley Regis, beside the Dudley Canal, where they established mines and, from 1848, the Old Hill blast furnaces. A second furnace was built by 1855. David Rose purchased Old Hill from Badger, but sold them on to Hingley as work on a new blast furnace at Moxley was finished.

Coal and ironstone was brought by tramway to the furnaces from the nearby Gawn and Old Hill (Blue Bell and Garretts Lane) collieries. Hingley also leased the Saltwells No.27 Colliery (Dudley Wood) from the Earl of Dudley. The Dudley Wood mines worked the coal and ironstone under the land formerly occupied by the Dudley Wood Furnaces, now demolished. By 1890, when Noah Hingley & Sons became a public company, the combined output of Hingley's mines was stated to be 3,000 tons per week.

The Netherton Ironworks was gradually extended along the Dudley Canal and came to occupy most of the north bank between Bishtons Bridge and Primrose Bridge. Part of the ironworks was erected on land reclaimed from the canal when it was widened and slightly straightened opposite the Chain and Anchor Works. The remaining section of the old waterway became an ideal basin and wharf for the loading of chains and anchors.

The Pelsall Ironworks comprised three separate ironworks. The arrangement of the puddling furnaces and mills are shown below in four sections of a large-scale map of Boaz Bloomer's Pelsall Ironworks.

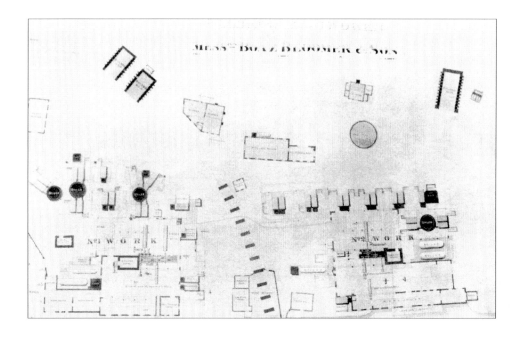

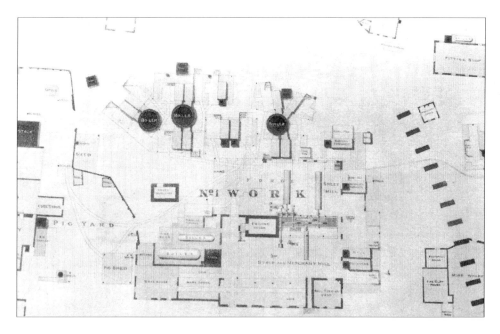

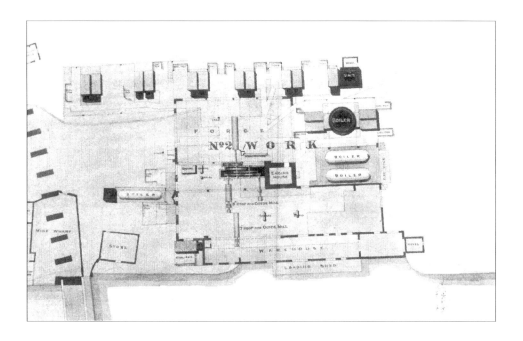

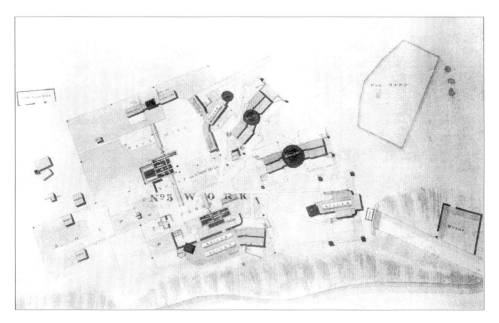

From Primrose Bridge, eastward to the Chain and Anchor Works, were buildings that were known as the 'New Side Ironworks' which comprised three forges. Within the confines of each forge were a number of ball, puddling and heating furnaces, merchant mills and forge trains. Wrought iron was produced here for the manufacture of their chains and anchors. It had to be of good quality for the chains not to break. Sheet and plate iron was also produced which Hingleys sold on to manufacturers and other ironworks.

The Chain and Anchor Works was the Old Side of the works. It included a Chain Shop, three end welding shops, two side welding shops, two anchor shops, an anchor forge, a shackle makers shop and a stud foundry. Each welding shop had a number of hearths where the chains were assembled, beside each hearth was a wrought-iron crane to assist the chain maker.

There was also an ironworks attached to the furnaces at Old Hill. This part of the works consisted of ball, heating and puddling furnaces, a forge train and a merchant mill. Standard-gauge railway sidings served both the furnaces and ironworks. Railway waggons were hauled up an incline from the Great Western Railway exchange sidings at Cox's Lane. Pig iron was obtained from a number of sources. Stock on hand in June 1890 included iron from Barrow, Lilleshall, Heyford, Holwell, Madeley Wood, Thornecliffe and Wellingborough. Most would have been conveyed to the nearest railhead and forward by boat.

It seems that iron was received from three railway companies. Traffic over the GWR lines would have ended up at Wythymoor and would have been carted, or boated, by Thomas Bantock the short distance into the works. That Hingley made regular payments to the Shropshire Union Co. for freight indicates that they handled iron traffic from the LNWR basins on the other side of the Netherton Tunnel. Ironworks served by the Midland Railway, such as Wellingborough, evidently sent iron to Great Bridge for transhipment into Midland Railway boats. There was another important traffic in finished iron from Netherton and Old Hill, which returned to the railway basins for transhipment. Chains and anchors were sent all over the country, and in particular to the shipbuilders on the Clyde and Barrow in Furnace.

Hingley produced plate and sheet iron to different thickness and lengths. There was a significant traffic in finished iron along the Dudley Canal to local ironworks. Samuel Lewis at Wythymoor received iron from Hingley and supplied nails in return, as part payment for the goods. George Hartshorn of the Atlas Iron Tube Works was another regular customer.

The years 1876 and 1877 marked another down trade for the South Staffordshire pig ironmakers. Cheap iron was coming into the district from other furnaces, which the local ironmasters were finding difficult to compete with. There seemed little prospect of improvement and as the years went by, more and more furnaces were shut down. Some responded by the temporary stoppage of furnaces. Willenhall Furnaces Ltd, for example, operated sometimes one and sometimes two furnaces. Other operators completely stopped production with disastrous effects for the workforce.

In order to stave off some of the worse effects, miners were asked to work longer hours for less pay, resulting in one group from Bilston stating they would rather eat grass rather than work a nine-hour day! Thorneycroft & Co., proprietor of the Shrubbery and Swan Garden Ironworks, informed his workers that unless they would accept a reduction in wages, the firm would have to close. On the advice of their unions, the men refused to do so. Thorneycrofts closed their ironworks and offered them for sale.

There were many factors which caused this state of affairs, exhaustion of minerals being the prime cause. By 1871, Thorneycroft, at their Bradley Furnace, were already importing all their ores and were buying in pig iron from a variety of sources. Thorneycrofts were not alone in this situation, many of the local ironmasters were increasingly relying on imported ore and several firms chose the option of closure and disposal of works.

CHAPTER 8

TWILIGHT YEARS OF THE IRONMASTERS ON THE ROAD TO DIVERSITY, 1881–1900

Demand for iron products became more diverse during the second part of the nineteenth century. Developments in civil and mechanical engineering generated the need for structural ironwork and improved mechanical parts. Precise measurement became a key factor, as did the quality of iron. Ironmakers came to produce an ever-increasing variety of grades of iron, suited for different purposes. There was also an increase in finished ironworks that brought in wrought iron to roll into sheets and plates. With diversification came better descriptions for the different trades. Those that smelted iron ores to make pig iron became known as 'pig ironmakers', while those that made pig iron malleable and produced bar iron were generally called 'ironmakers'.

Mechanical Puddling

Mechanisation was playing an increasing part in the trade and with typical Victorian ingenuity steam power was applied wherever possible. Applications included steam cranes, steam hammers and steam-puddling mechanisms. Writers and researchers have frequently commented on the lack of success achieved with steam puddling. Nevertheless, several firms decided to adopt the technique.

During 1864 the firm of Ambrose, Beard & Sons, Regents Ironworks, adopted the use of steam-puddling furnaces that were designed by the inventor, Mr Griffiths. The machinery consisted of a rabble bit (similar to the conventional puddler's iron bar) which was moved by machinery worked by steam. The iron was stirred in the same fashion as with the manual method of stirring. When the boiling ended and the iron was starting to coagulate, the machinery was stopped and the puddler (who had been saved the hard work of stirring) began to divide the now coagulated contents of his furnace into balls of suitable size for shingling.

Morewood & Co. also adopted steam puddling at their new works at Woodford, near Smethwick. These works were erected during 1876 to replace their existing operation at Ettingshall. Morewoods had taken over the Ettingshall Ironworks that had been established by Thomas Banks & Sons in the 1820s. By 1876 these were considered wasteful on fuel and the new Woodford Works were put together with economy in mind. They originally installed a furnace designed by Casson, but went on to remodel the plant to their own special designs. As with the Griffiths process, a steam-powered rabble bit was used to stir the metal.

Furnace and Ironworks Closures

Those who worked up iron found, through improved railway communication, a greater variety of basic metal was available from blast furnaces in different parts of the country. Local blast furnace owners faced stiffer and stiffer competition. Another important factor was the general exhaustion of local coal and ironstone seams. Some furnace owners chose to import ironstone from elsewhere

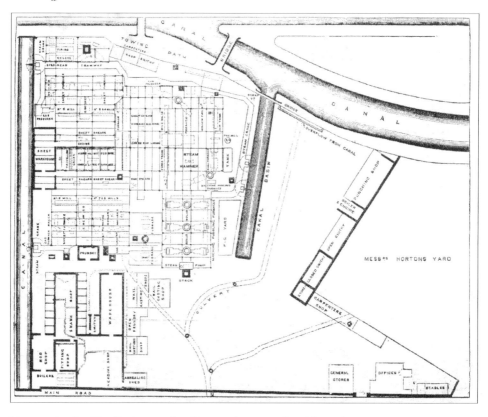

Morewood & Co., Woodford Ironworks. This 1882 plan is reproduced from *The Engineer*.

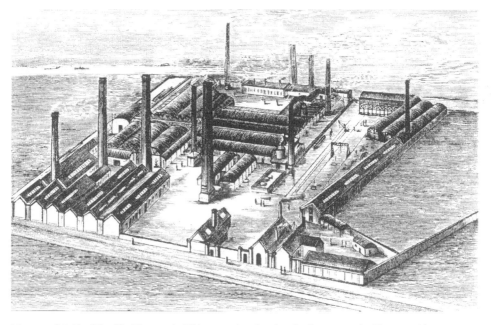

Morewood & Co., Woodford Ironworks. This engraving showing the furnaces and mills covered by corrugated-iron roofing.

Benjamin Whitehouse
(*Edgbastonia*).

and drew on the new, deep coalmines that were being developed on Cannock Chase. Other ironmasters ceased furnace operation either through bankruptcy or their own choice. Another slump in the iron trade, between 1875 and 1886, seriously affected the Black Country iron industry and in 1890 only twenty-seven locations remained where pig iron was made.

Declining local coal supplies had an adverse affect on some ironmasters. The firm of H.B. Whitehouse & Son endured failure in 1878. Henry Bickerton Whitehouse had died, leaving his son, Benjamin Whitehouse, in sole charge and it was Benjamin who faced the creditors alone. The coal and ironstone mines around the furnaces at Priorsfield and at New Cross, Wednesfield, were greatly depleted and calcined ore received from North Staffordshire supplemented the supply to the furnace bank. New Cross Colliery was sold of and creditors accepted a reduced payment for their debts. Thus the firm of H.B. Whitehouse survived; other ironmasters would be less lucky.

Casualties of this period included the Addenbrookes (Rough Hay), John Bagnall & Sons (Caponfield & Goldsgreen), Corbyns Hall Furnaces, Chillington Iron Co. Ltd (Chillington & Bentley), David Jones (Herberts Park), Parkfield Furnaces, W. & J.S. Sparrow (Stow Heath & Millfields), W.H. Sparrow (Osier Bed) and Willenhall Furnaces Ltd (Willenhall).

Willenhall Furnaces closed in 1882 when most of their resources were exhausted. The blast furnace plant, as described in September 1881, had seen many changes since first constructed, in order to keep in touch with the latest improvements. The three furnaces were now 55ft high and fitted with gas apparatus that collected the blast furnace gases which were used for heating the stoves and generating steam in the boilers. There were five hot-air stoves. The blast engine house contained a condensing beam engine, powered by steam which was supplied by six egg-ended boilers. There was a wrought-iron incline from the calcining kilns to the top of the furnaces. The calcining kilns were two in number, constructed of brick and cased in iron, one 45ft high, the other 43ft high. There were sixteen coking hearths in typical Black Country practice that coked coal in the open air.

Neal Solly and his brother, James, had both decided to make Edgbaston, Birmingham, their home. For Neal, whose previous homes had included the imposing Moseley Hall at Bushbury,

Featherstone, Westfield Road was modest by comparison. James moved to Frederick Road and was still associated with the Great Bridge Iron and Steel Co. when he died in 1889. For Neal, the closure of Willenhall Furnaces was a release from the role of managing the furnaces and gave him an opportunity for retirement. He was now free to follow his passion for painting. He spent his remaining years as an artist and also found time to write a biography of local Edgbaston artist, David Cox. Neal Solly died on 8 August 1895.

Rough Hay Furnaces closed after the tragic death of John Addenbrooke at Wednesbury. On 15 February 1882 John had visited his colliery at Leabrook, which lay close to the Great Western Railway Interchange Basin sidings. Addenbrooke was a commissioner of the South Staffordshire Mines Drainage Commission and was anxious to attend their meeting at Wolverhampton that afternoon. Instead of going to Wednesbury Station along the road, John chose to follow the railway tracks. Walking along the side of the 'down' track, that is, the line of tracks used by trains running between Birmingham (Snow Hill) to Wolverhampton, to the Great Western Station at Wednesbury, John Addenbrooke was run down by a train and killed. Rough Hay Furnaces were subsequently advertised for sale and dismantled. Part of the plant was transferred, including perhaps one of the steam locomotives, to Wellingborough in order to assist with the making of the new ironworks there.

After standing idle for a time, the Parkfield Works were taken over by John and Thomas Williams, who invested new money in the plant to restart the furnaces during 1881. They now became known as the 'Old Park Furnaces'. It was a brief respite. Ryland's *Directory* reported that all five Parkfield Furnaces were pulled down after 30 June 1888.

During January 1889 the sale of Corbyns Hall Furnaces was announced. These premises were owned by the trustees of the late Benjamin Gibbons and were formerly occupied by William Mathews. Three iron-cased blast furnaces, 45ft high, and one brick-built furnace, incline engine and two blast lifts remained.

One of the greatest losses was the closure of the Chillington concern. The Barkers had built up the concern, adding mines and ironworks to the original ironworks established in 1829. By 1878 many of the mines had been worked out and the principal mining activities were conducted at Bentley. With deteriorating property, a new direction was sought within the business of edge tool manufacture. A brand new works was laid out on land north of the furnaces for making edge tools.

Alfred Hickman gained a shareholding in the Chillington Co. and, as debts rose, provided a mortgage for the works. During 1884 it was decided to sell off the Bentley Colliery. Then, in December 1884, a decision had been made to close the ironworks and furnaces at Chillington, place the company in receivership and dispose of their other works. Workers were laid off and there was a general concern for the unemployed. Disposal was made in a piecemeal fashion between 1885 and 1887. Chillington Ironworks and blast furnaces were demolished and Leabrook and Capponfield Ironworks were sold. Leabrook was finally disposed of during 1887 and converted into a tube works. The Edge Tool Works was also sold off to an independent company, while Hickman retained the land. He was the principal creditor. It was this control that enabled Hickman to arrange with the London & North Western Railway to take the former canal basin and boat dock for the Chillington Works and adapt it as a railway interchange basin (1898–1901).

Capponfield Ironworks had a long history in the iron trade, including ownership by the Bishton family. James Foster and the Chillington Iron Co. had made improvements during their ownership of the plant. William Molineaux & Co. purchased these works in March 1885 for a sum stated to be between £7,000 and £8,000. The plant included a forge, three sheet mills and a hoop mill. Bars, billets, strip and sheet were then manufactured at Capponfield.

Caponfield Furnaces, formerly in the ownership of John Bagnall & Sons Ltd, were sold to Alfred Hickman who, in turn, let it to Thomas and Isaac Bradley, then trading as the Bilston Brook Co., who rebuilt and improved them. During 1882 Darlaston Green Furnaces, iron and steelworks were offered for sale. This was an extensive site that occupied property on both sides of the Walsall Canal.

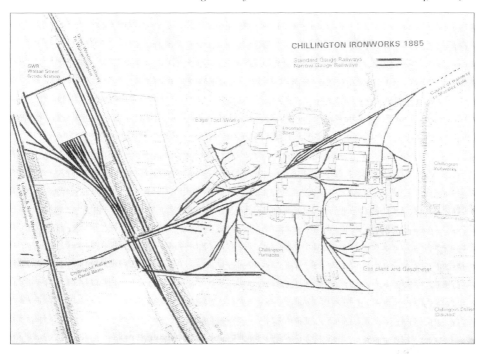

Map showing the Chillington Ironworks at the time of closure – based on the Ordnance Survey plan.

There were three furnaces, one near the iron and steelworks, the other two were on the opposite side of the canal. The pair of iron-bound brick blast furnaces with charging stage, lift and hot-air oven, 43ft long by 18ft wide and 23ft high, were leased from 1883 by T. & I. Bradley of Bilston and ironmaking continued. The third and oldest furnace was demolished while the ironworks was later worked by Tolley, Sons & Bostock.

Alfred Hickman was arguably one of the greatest Staffordshire ironmasters. He was the son of George Rushbury Hickman and Mary Haden (the daughter of Benjamin Haden of the Old Hall at Tipton). George was appointed manager of the Moat Colliery at Tipton, which was an extensive estate of mines placed alongside the Old Birmingham Canal and the Ocker Hill Branch. They had three sons, George, Edwin and Alfred. The brothers grew up at Moat and became well versed in the coal and iron trades. There were also two daughters, Frances and Sarah Haden. Sarah married Reverend Noot, the vicar of St John's, Dudley. Edwin (born 1824) appears to have died young.

After a grammar school education at King Edward's, Birmingham, Alfred followed the profession of an iron merchant and also became a partner with his father and brother at Bilston Brook Furnaces, and with just his brother at Groveland Furnaces and Stonefield Furnaces and Ironworks. Alfred and George survived a call for receivership in 1861, through the part disposal of their property. Stonefield Furnace and Ironworks was leased to two separate undertakings, while George Haden Hickman continued to work Groveland on his own account. Bilston Brook Furnaces, which had been originally belonged to the Price family, passed to Benjamin Gibbons junior.

The legal action that accompanied the creditor action took around two years to resolve and led to the two brothers embarking on different business ventures. Alfred chose to take over the Bilston Furnaces that originally had been established by the Bickley and Gibbons families and had been worked subsequently by the Sparrow and then the Jones family. Alfred Hickman made a success of the venture and soon made arrangements for linking the furnaces by railway sidings to the London & North Western Railway, enabling deliveries of minerals and despatch of pig iron by rail. He was associated with funding the pioneering work in basic steel manufacture that was conducted at the Patent Shaft Work, Wednesbury. The success of these experiments led to the building of the Staffordshire Steel and Ingot Works on land adjacent to the Bilston Furnaces and an increase of the company private rail system that served both works.

Alfred Hickman (1830–1910) was a keen promoter of both canals and railways and regularly campaigned for cheaper rates of carriage. As chairman of the Wolverhampton Chamber of Commerce, 1883–1884, he made a serious attempt to reduce railway rates for goods in the Midlands. His views on local industry propelled him into politics, where he stood as Conservative candidate for Wolverhampton, eventually winning a seat and starting a long a parliamentary career. Alfred was MP for Wolverhampton West from 1885–1886 and 1892–1906. His industrial holdings increased while other ironworks were closing. In addition to his Springvale Furnaces and Staffordshire Steelworks, Hickman was connected with Haunchwood Colliery, near Nuneaton,

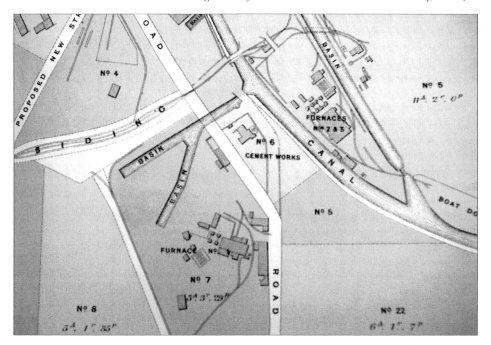

Darlaston Steel and Ironworks sales plan, showing the location of the furnaces, 1882.

Right: Darlaston Steel and Ironworks, showing location of the steelworks and forges.

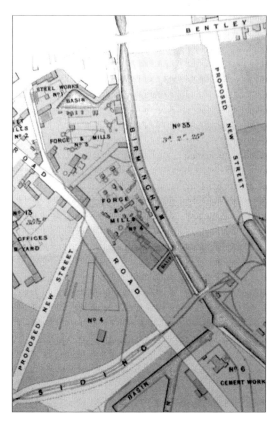

Opposite: Caponfield Furnaces, as seen from the Old Main Line. The furnaces were of the iron-cased type, as was commonly adopted during the second half of the nineteenth century.

Alfred Hickman.

Postcard view showing the pig beds of Spring Vale Furnaces and one of the Barclays locomotives employed there.

Inside Bankfield Ironworks, 1899.

Mill No.2, Bankfield Ironworks.

George Thorneycroft.

David Rose & Sons survived a bankruptcy petition and ironmaking continued through to 1885. Eventually the furnaces were blown out and the works, plant and locomotives offered for sale during August 1886.

One of the largest finished iron factories belonged to the firm of George Benjamin Thorneycroft & Co. George had gained his experience working for the Addenbrooke family at their Moorcroft Furnaces. He then started the Shrubbery Ironworks at Wolverhampton, beside the BCN Main Line. Thorneycroft sought to gain as much business as possible here from the growing railway industry. At Shrubbery, rails were made from the best charcoal iron. Patent axles and railway tyres were also produced at the works.

Later acquisitions included the Swan Garden Works at Wolverhampton and the Hatherton Furnaces at Bloxwich. In January 1860, G.B. Thorneycroft & Co. advertised Hatherton for sale; they had purchased the Bradley Colliery which had once belonged to John Wilkinson. Two new blast furnaces were erected at Bradley and had commenced to make pig iron by 1862. Pig iron made at Bradley was conveyed by boat to the Wolverhampton Ironworks for finishing. Thorneycroft's had a good name for plate and best iron of all kinds. They also had a reputation for producing what was termed 'bullet iron'. This type of iron was prized in Manchester for the manufacture of parts for cotton machinery.

After George Thorneycroft's death the firm was managed by a group of businessmen that included his only son, Major Thomas Thorneycroft, and John Hartley, his son-in-law. Profits dwindled during the 1870s. The managing partners showed good business sense and decided to shut all the works down. Closure came in 1877 when the firm was still in profit. After lying idle for a few years, the Shrubbery Works were advertised for sale. The plant included sixty puddling, scrap pile and mill furnaces, fourteen forge and mill trains, a steam hammer and eight condensing beam engines, ranging in power from 40-120HP. In 1881, Shrubbery Ironworks were stated to have passed to Nettlefold's of Birmingham, who had bought it to make sheet iron. In April 1886 the *Colliery Guardian* reported a new future for the site – one that favoured rail transport:

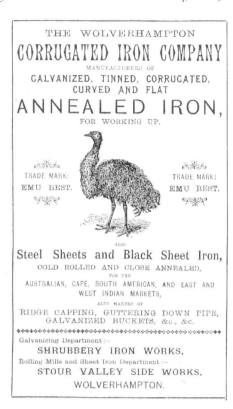

Wolverhampton Corrugated
Iron Co. advertisement.

Lysaght's rolling mill at Swan Garden Ironworks, Wolverhampton.

A steam hammer at Lysaght's Swan Garden Ironworks.

Rapid progress is being made with the new works of the Shrubbery Iron & Steel Co. which is being erected on the site of the iron making firm of G.B. Thorneycroft. The works are 6 acres of land and railway siding has been put in so that the company will have both rail and canal access. The sidings are so situated that the company will be able to unload pigs almost right into the puddling furnaces.

The Shrubbery Iron & Steelworks occupied only part of the original ironworks site. The Wolverhampton Corrugated Iron Co. was to take the other (Stour Valley) side. Corrugated iron was an important development for the local industry. Its uses rapidly spread to include roofing and fencing. The Wolverhampton Corrugated Iron Co. (run by John and Joseph Jones) was an established firm in Church Lane and Dudley Street, Wolverhampton, when they took over the Stour Valley side works. They gained better transport facilities which, in turn, favoured their business.

Another John Jones became head of the company and together with his brother, Joseph, and Edward Farnworth helped build up the concern. In 1893 both works, Shrubbery and Stour Valley, were united under the name of the Wolverhampton Corrugated Iron Co. John Jones lived at Ireton Lodge, Goldthorne Hill. He was a magistrate and also a councillor for St George's Ward from 1872. John was Mayor of Wolverhampton for three years in succession. He died in 1897. Meanwhile, the Wolverhampton Corrugated Iron Co. continued to expand. In 1905 they opened a more extensive plant at Ellesmere Port that was served by both rail and canal.

At Shrubbery most of the Wolverhampton Corrugated Iron Co.'s iron was despatched by rail, but coal came in by canal boat. In around 1910 the decision was taken to close the Wolverhampton works and concentrate their business on Ellesmere Port. Knowles' Oxygen Co. took over the Shrubbery Works site, while the Stour Valley Works were demolished.

Thorneycroft's other ironworks, the Swan Garden Works, lay beside the Wyrley & Essington Canal at Wolverhampton. They were purchased by John Lysaght, sheet ironmakers of Bristol, in 1878.

Edward Bagnall Thorneycroft was less fortunate with the iron trade. As the son of Edward Thorneycroft and Martha Bagnall, he was christened at St Lawrence church, Darlaston, in 1818. Edward grew up to follow his father's and uncle's occupation. He went into partnership with Samuel Griffiths and came to operate the Staffordshire Ironworks at Greets Green.

The Staffordshire Ironworks were improved by Griffith & Thorneycroft through the putting down of a third rolling mill. The works then came into the possession of Sir Francis Goodrike and were briefly tenanted by Thomas Cross, whose bankruptcy was announced in 1863. The Ironworks were then disposed of to W.C. Standish, a Southampton businessman, who traded as the Eagle Coal & Iron Co. A considerable reputation was built up for the brand of iron made at Greets Green. The Eagle Works were adjacent. These works were older and have a history that can be traced back to Thomas Price, when it was known as 'Bromwich Forge'. The arrangement of puddling furnaces and mills was haphazard, in common with many other contemporary Black Country ironworks. Working of the Eagle and Staffordshire Ironworks was passed onto the Jones brothers, who supplied them with iron made at their Buffery and Green Lane Furnaces.

Eagle Works subsequently passed to Ebeneezer Parkes junior MP and a Justice of the Peace of Edgbaston, Birmingham. Ebeneezer Parkes senior had carried on the iron trade at the Atlas Works in West Bromwich for many years and he was succeeded in the management by his son, Ebeneezer. Another son, John Parkes, carried on the Birmingham Corrugated Iron Co.

Sourcing Fresh Ironstone Supplies by Canal and Railway

Local ironstone deposits had been virtually exhausted by 1880 and firms had to get their ironstone from other sources in order to survive. Those that continued to trade had the choice of receiving ironstone by rail or canal. Both options remained open, but there was a growing tendency for

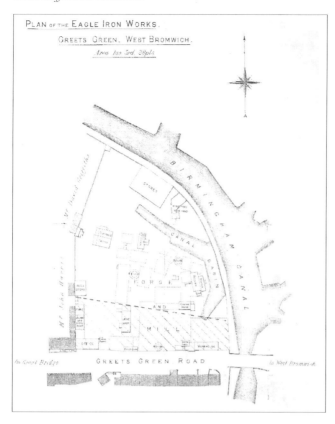

PLAN of the EAGLE IRON WORKS.
GREETS GREEN, WEST BROMWICH.

Eagle Ironworks sales plan, 1885. These were of the type developed during the 'dry puddling' era and were previously known as West Bromwich Forge. The mills and furnaces were arranged in a haphazard manner across the site.

Ebeneezer Parkes, Junior MP and proprietor of the Roway and Eagle Ironworks. (*Edgbastonia*)

firms to change from canal to rail after 1870. Sidings were built and connected with the main line railway companies. Many works developed extensive internal systems and used their own locomotives to handle the traffic.

Several ironmasters chose to establish their own ironstone quarries. Alfred Hickman had quarries in North Wales and East Adderbury, Oxfordshire. The New British Iron Co. had mines near Hook Norton in Oxfordshire, while Willingsworth Iron Co. leased mines at Burton Dassett, Warwickshire.

Willingsworth Iron Furnaces retained a fleet of boats until the furnaces closed. It was once a large fleet and, as late as 1907, the quarterly BCN tonnage bills would exceed £500. Railway sidings had been made into the furnace yard by 1880, but canal traffic evidently remained high. Account books survive from 1905. The normal traffic by canal and rail then included cinders, fuel, ironstone and limestone. There were also regular deliveries of manganese ore. The lease for the Burton Dassett quarries had commenced in December 1905 and ironstone started to arrive by rail from these quarries in 1907. This new ironstone traffic coincided with improvements and alterations to the blast furnace plant and railway sidings at Willingsworth.

Five of the surviving blast furnace sites at BCN, Brades, Brades Hall, Coseley Moor, Hange and Lays never had railway sidings. All traffic was either boated or carted to and from their works. Minerals were often collected in their own boats or sent by a carrier. There was also some traffic from the nearest railhead.

John Round's boats collected coke, ironstone and limestone. Traffic figures along the Staffordshire & Worcestershire Canal for the period 1884–1889 show a regular traffic in Staffordshire ironstone. An average of thirty boatloads per month passed from Great Haywood to Aldersley, carrying ironstone and some coke for the Hange Furnaces. W. & G. Firmstone, who had the Lays Furnaces on the Stourbridge Canal, also had a regular trade in ironstone and coke to Stewponey.

Coal and coke continued to be moved by boat to the other surviving furnaces. Despite a rising demand for steel, iron continued to be a staple product of the Black Country. Firms such as G. & R. Thomas at Bloxwich were successful manufacturers, when others failed. Their furnaces produced pig iron well into the twentieth century. Hatherton Furnaces were established beside the Wyrley & Essington Canal by William Fleeming Fryer, and owners included G.B. Thorneycroft before the tenure of George & Richard Thomas.

George and Richard Thomas operated a fleet of day boats, which carried material all over the West Midlands. Accounts held in Walsall Record Office show that the canal remained an important means of transport. There were regular movements of boats bringing coal and slack from their collieries at Wood Farm and Fishley to their furnaces. Sand for the pig beds came from Ogley Hay and there was also some traffic in calcined pyrites to Hatherton. Ironstone was also brought along the Wyrley & Essington Canal to the furnaces from the Deans Colliery, near Wolverhampton.

Finished iron, including bars, hoops and strips, was produced at the Thomas's Birchills Ironworks. Each week, 140 tons of finished iron were made at the works, which had formerly belonged to James Bissell & Son. Pig iron was taken by boat to various canalside ironworks such as the Cleveland, Mars and Swan Garden Ironworks at Wolverhampton and to other works beside the Walsall Canal.

Canal traffic in ironstone from Adderbury continued to dwindle after 1881 when the tramway had been extended to sidings beside the GWR. Some trade did remain and, in August 1887, iron oxide was granted the same rate as ironstone on the Warwick & Napton Canal. In 1889 Alfred Hickman of the Spring Vale Furnaces, Bilston, took over the East Adderbury Quarries.

Some ironmasters continued to collect ironstone by boat for as long as they remained in business. Round Brothers, of Hange Furnaces at Tividale, possessed a number of cabin boats which travelled to Staffordshire and along the Shropshire Union in order to bring stone to their furnaces. The iron ore, when transported by canal or rail, was usually calcined at source.

Another small source of iron was obtained from iron pyrites, which were often brought into the midlands via the Port of Gloucester. Pyrites were used in the chemical industry to manufacture

This page: William Roberts worked the Tipton Green Furnaces at Tipton, which were originally built in 1808 and served by canal. Railway sidings were eventually made to bring materials to the furnaces, move pig iron and take slag to a designated tipping area. The link made with the London & North Western Railway involved tight curves. Roberts, like other contemporary South Staffordshire Ironmasters, employed locomotives to haul wagons on their private tracks. During 1892 the sidings and private lines were enlarged and a new locomotive was ordered from W.G. Bagnall of Stafford.

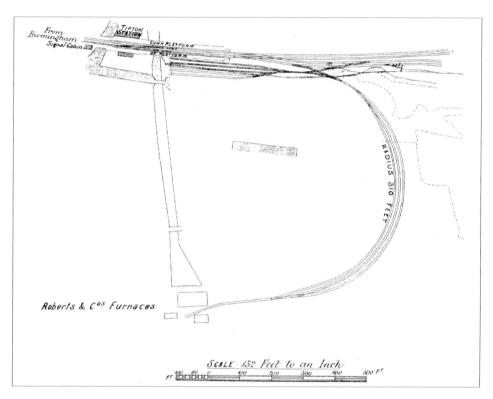

Robert's sidings at Tipton, 1890.

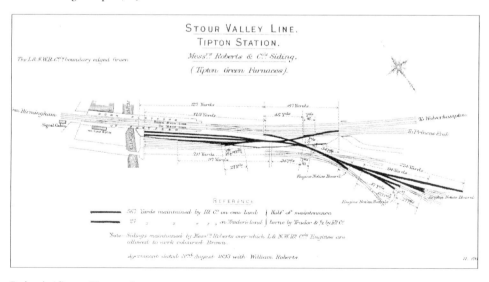

Robert's sidings at Tipton, 1893.

sulphuric acid and the 'waste' ore was disposed off to the local furnaces to extract the iron. The term applied to the waste ore was 'purple ore'. Transport of purple ore was made by canal boats or railway wagons from the chemical works to the ironworks. Firms such as J. Bagnall & Sons and Noah Hingley used purple ore as part of the puddling process.

Despite the falling number of furnaces, red ore traffic continued from Lancashire and Cumberland. In August 1878, W. & J.S. Sparrow applied for, and was granted, a drawback on tolls for red ore passing along the Shropshire Union and Birmingham Canals from Ellesmere Port to Millfields Furnaces. There were extensive iron ore docks at Ellesmere Port, where carriers could collect by boat. W.S. Foster of Tipton was engaged in this trade until at least 1900. Although less ironstone and iron ore was taken by canal boat after 1880, furnaces closed or switched to rail transport.

Furnace and Ironworks Closures

A long association with ironmaking ended when Henry Onions Firmstone closed down his operations at Crookhay, the Hyde, near Kinver. Crookhay Furnaces were sold in December 1888. Four blast furnaces, two iron-cased and two brick-built, were listed, along with the blast engine and gas tubing that collected the gases from two furnaces to supply the hot-air oven. Lays Furnaces were mentioned in a dismantling sale of October 1895 that included the materials of three blast furnaces and two blast engines made by Holcroft of Bilston.

Increased competition from ironworks and pig makers elsewhere in the country took its toll on the local manufacturers. Mild steel was also replacing iron in many products. The survivors of the iron trade were those that adapted to serve the changing needs of society. These were the makers that supplied the engineering parts for the gas and gas engine industries, structural iron and steelwork for bridges and buildings, pipes for the water industries, railway tracks, galvanised roofing, plated wares, cased tubes for the metallic bedstead trades, and most importantly the new industry of weldless tubes that was necessary for the production of bicycles.

Pelsall Coal and Ironworks, which had made such a successful start, faced tough times during the 1880s. Despite the sinking of new mines the trade suffered and directors finally decided to close the ironworks during 1891. The contents of the three ironworks was sold off during 1892 and then they were dismantled. Some of the mines were sold to the neighbouring Walsall Wood Colliery Co., who worked them until 1903.

The New British Iron Co. faced similar problems and their investment in two Siemens Martin steelmaking furnaces came too late to stave off administration. Attempts to sell off the iron and steelworks and mines at Corngreaves was made in 1893 and again in June 1894. No offer was forthcoming and the extensive property was sold off piecemeal between 1894 and 1897. The estate then included the Corngreaves Iron and Steelworks, six blast furnaces, a chain works, testing house, saw mills, six collieries, railways and rolling stock and Corngreaves Hall. The whole business was still at work, being carried on by the receivers. The sale attracted few bids. Sir Benjamin Hingley (N. Hingley & Sons) offered £25,000 for the whole estate, but this was not enough and the auctioneer went on to offer individual lots. Only a few plots of land were actually sold. Another set of sales was then arranged. The stock of iron and steel was put up for sale in September and November 1894, including steel ingots.

In June 1895 the Siemens Martin Furnaces, which made steel by the acid process, were advertised for sale. In July 1895, the building formerly used as the chain works, testing house and timber yard was also put up for sale. During September 1895 the material of three blast furnaces, a furnace hoist and parts of the ironworks with annealing furnaces and a bull dog mill were for sale, while in October some of the standard-gauge and 3ft 2½in-gauge rolling stock was disposed of. The sales in 1896 included the 20in forge train, merchant and strip mills and a steam hammer. Codsall Colliery was offered for sale in 1896 and the Greenfields Colliery at Willenhall in 1897. The final sale was made in April 1897 when Corngreaves Hall was sold.

CAUTION.—In consequence of the numerous Frauds which have been committed by Chain Cable Manufacturers, in accepting contracts, both from users and foreign buyers, for the supply of Chains to be made of "LION" IRON, but using instead an Iron of inferior quality, afterwards testing the chain at a private testing machine, instead of submitting it to the impartial Test of Lloyd's Public Testing Machines, thus defrauding the buyer and throwing upon him risks of subsequent failures and breakages in collieries, factories, ships, &c, &c.

We, the undersigned Company, who have since the year 1825 been the Sole Makers of the brand of Iron known in all markets as "LION," consider it incumbent upon us to issue a Public Notice that we shall in future PROSECUTE ALL PERSONS who are found to be Fraudulently Representing Chains, Cables, or other Iron Goods made of an inferior iron as having been made of "Lion" Iron.

Whilst we think it our duty to publish the following Apology which we have received, we desire it to be clearly understood that in future we shall accept no Apologies, but shall in each case that comes under our notice after this date Prosecute the Offender according to law.

THE NEW BRITISH IRON COMPANY,
LIMITED.

Corngreaves Ironworks,
near Birmingham.

———

[COPY.]

To THE NEW BRITISH IRON COMPANY, LIMITED,
Corngreaves, near Birmingham.

We the undersigned firm of chain cable makers hereby acknowledge and declare that on the 6th day of July last we received a written order for three iron chains, together 1,600 yards in length, to be made of "Lion" Iron, which we well knew was your special brand, and had been for many years, and that you exclusively manufacture such brand at your Corngreaves Works, which are distant two miles from our place of business.

And we further acknowledge and declare that we caused such chains to be made of an inferior iron to "Lion" Iron, and that the certificate forwarded with the chains emanated from the manufacturers' own works.

For these acts we humbly beg the forgiveness of The New British Iron Company (Limited), and undertake not to offend in like manner again.

Dated this 24th day of December, 1885.

(Signed)   WM. INGLEY & CO.,
803   Quarry Bank, near Brierley Hill.

New British Iron Co.'s notice concerning their quality and brand of iron.

174

A much reduced operation was left. The two most modern furnaces were taken over by the Corngreaves Furnace Co., who continued to work the furnaces as well as the Fly and Black Wagon Collieries. The New Hawne and Timbetree Collieries were taken over by Shelagh Garrett. Other mines, such as Codsall, were closed.

The number of blast furnaces continued to decline. By March 1884, 118 remained but only forty were in blast. In effect only a third of the South Staffordshire furnaces were in use. It was in fact one of the lowest figures, to be rivalled perhaps by Shropshire which had only six out of twenty-three at work (23 per cent). High performers included Lancashire with thirty-two out of forty-nine, North Yorkshire with seventy-seven out of 100 and Scotland with ninety-three out of 148.

Steel Making

South Staffordshire Iron was not suitable for steel making, mainly due to elements such as phosphorus being present. Those who chose to make steel in the Black Country and Birmingham did so by importing iron into the district to make crucible steels for edge tools, weapons and wire. Later, with the Bessemer process, mild steels made at works such as the Patent Shaft & Axletree at Wednesbury proved important to the manufacture of railway rolling stock.

A lifeline was thrown to certain ironmasters through the manufacture of steel by the basic process. The technology for this method was developed during the 1880s and led to the building of the Staffordshire Steel & Ingot Works at Bilston, adjacent to Springvale Furnaces. The method was specially adapted to remove phosphorus from the iron made in South Staffordshire and render it suitable for steel making. A profitable byproduct was the phosphorus rich slag that was sold as a fertilizer.

Ironworks and Furnace Survivors

Between 1885 and 1900 South Staffordshire Furnaces (and the ironworks they supplied) continued to cease operation and were pulled down. The surviving iron furnace owners improved production through rebuilding and improving their remaining furnaces. They also produced a range of different pig irons to suit local needs. An increasing number of furnaces came to be operated by corporate companies, although a few of the established ironmaster firms doggedly continued to trade. J.H. Pearson of the Farcroft, Handsworth, was one of these independent ironmasters. He operated the Netherton and Windmill End Furnaces, which were served by the Dudley Canal. Netherton also had a siding connection with the Great Western Railway at Wythmoor. J.H. Pearson also owned Parkhouse Colliery, near Chatterley, that had sidings to the North Staffordshire Railway.

Cochrane & Co. at Woodside owned the Woodside Ironworks and also the colliery at Yew Tree, near Old Hill. As Cochrane & Co. Ltd, the Cochrane family were associated with the Ormesby Ironworks, Middlesborough, and four blast furnaces. They had New Brancepeth Colliery and ironstone mines at Saltburn and Guisborough.

Philip Williams & Sons operated their original iron furnaces and works at Wednesbury Oak, making alterations and improvements as required. Their forges and rolling mills turned out angles, bars, sheets and plates, much as they had done so for previous eighty or so years.

With the new century, furnace numbers were close to the figures of 100 years before. The main difference was that production at a modern furnace was far greater than in those existing in 1801. Ryland's *Trade Directory* recorded forty furnaces, operated by fourteen separate undertakings. There were also another two furnaces at Green Lane, Walsall which, at that time, may not have been at work.

Blast furnace workers at Corngreaves Furnaces.

A view of the pig yard at Corngreaves Furnace, which shows one of the narrow-gauge locomotives on the private railway there.

Ryland's *Trade Directory,* 1902 – List of Pig Ironmakers

Proprietor	Furnace Name	No.	Pig Iron Produced	Brands
Thomas & Isaac Bradley Ltd	Caponfield	3	Mine, Part Mine, Cinder	Three Brands
Thomas & Isaac Bradley & Sons	Darlaston Green	2	Part Mine and All mine	IXL
Cochrane & Co.	Woodside	2	Mine	Woodside, C. & C.
Earl of Dudley	New Level	5	Cold Blast, All Mine, Hot Air, Part Mine	LNF, DUD, LNF-XX
Robert Fellows	Corngreaves	2	Cold Blast Mine, Part Mine, Common Pig	Corngreaves, Lion
M. & W. Grazebrook	Netherton			
Alfred Hickman Ltd	Spring Vale	6	Mine & Basic Hydrate, Part Mine	B.F.M.S.V.H .
N. Hingley & Sons Ltd	Old Hill	2	Mine	Old Hill
J.H. Pearson	Netherton	2	Part Mine	Netherton
J.H. Pearson	Windmill End	2	Cinder	Windmill End
W. Roberts (Tipton) Ltd	Tipton Green	3	Part Mine, Basic Cinder	Roberts Tipton Green
George & Richard Thomas	Hatherton	2	Mine	G. & R.T., Special Hatherton, Wall End
H.B. Whitehouse & Son Ltd	Prior Field	3	Mine, Part Mine	Priorfield, Walbrook
Phillip Williams & Sons	Wednesbury Oak	2	Cold Blast, Hot Blast	W.O., Oak, P.W.
Willingsworth Iron Co.	Willingsworth	2	Basic	Willingsworth

Total Furnaces   40

Public companies, with boards of directors and shareholders, were a new force in the iron trade. These groups often had a variety of interests in which ironmaking and working only formed a part. This was a far cry from the days of the ironmasters, where the partners that set up the operations and risked their capital in ventures to gain wealth or debt. The company system provided better protection for the owners, but the responsibility of the board of directors was often diverse and geared for profit. In times of poor trade the tendency was to close the ironmaking operations.

These times were far removed from the days when the iron trade was still in its formative state. Yet in 400 years, from the seventeenth century to the twentieth century, local supplies of ironstone were nearing exhaustion. The annual local ironstone production decreased to a mere 38,000 tons by 1900 – the days of the ironmaster were clearly numbered. Those that still practiced the trade did so by bringing ore in by rail, paying the cost of transport. There was still a justification for firms to continue trading which was basically down to the supply chain that had been established for other related industries, such as the manufacture of automobiles, cycles, metallic bedsteads, structural ironwork and tubes.

With the decade commencing in 1890 came the twilight years of the Black Country iron trade, as well the declining influence of the independent South Staffordshire ironmasters. They were a breed of men that as a group contributed so much, sometimes at great personal sacrifice. Their legacy is the surviving features of their age, such as the bridges and the buildings. Another legacy is their names, handed down in historical documents praising their contribution and influence on

Above and below:
Two views
of Priorsfield
Furnaces, *c.*1910.
The image left
shows the lift
that carried the
charge to the top.
As Staffordshire
furnaces increased
in height, inclines
were frequently
used to carry the
material to the top.

the trade. Perhaps the most influential were Dud Dudley, Richard Foley, Abraham Darby, Richard Knight, John Wilkinson, Samuel Fereday, Richard Smith, John Bagnall, John Gibbons, John Hall, James Foster, Simon Blackwell and Alfred Hickman.

Sadly, not one iron furnace has survived and few buildings remain that can trace their existence to the days of the puddling furnace and rolling mill. West Midlands industry moved on to supply the automotive and aerospace trades through the modern methods of working metal by extrusion and pressing. Despite all these advances, however, there was still a need for castings and some Black Country Iron foundries survived to meet these demands. But now, with automotive production in decline, these last bastions of the iron trade are closing. A dogged survivor is the Aga Rayburn Foundry at Coalbrookdale, which still supplies cast-iron parts, continuing the tradition the Darby family established in 1708.

APPENDIX ONE

CHARCOAL BLAST FURNACES IN SOUTH STAFFORDSHIRE, EAST WORCESTERSHIRE AND NORTH WARWICKSHIRE

Furnace	River
Aston	Hockley Brook
Beaudesert	Shropshire Brook (tributary of Trent)
Bromwich	Tame
Cannock (Lower)	Rising Brook (tributary of Trent)
Cannock (Upper)	Rising Brook (tributary of Trent)
Cheslyn Hay	Wyrley Brook
Compton	Stour
Coven	Wyrley Brook
Cradley	Stour
Coneygre (Dudley)	Tributary of Tame
Deepmore	Wyrley Brook
Glasshampton	Dick Brook
Grange	Smestow Brook
Hales	Stour
Himley	Holbeache Brook
Middleton	Tributary of Tame
Perry	Tame
Poole Bank	Bourne (near Shustoke)
Rushall	Tame
Teddesley Hay	Tributary of Penk (near Penkridge)
Wolseley	Stafford Brook (tributary of Trent)

Dud Dudley's Furnaces

According to his own account, iron ores were smelted with coal at the following locations:

Furnace	River
Cradley	Stour
Hasco Bridge	Holbeache Brook
Himley	Holbeache Brook

APPENDIX TWO

WATER-POWERED FORGES FOR SOUTH STAFFORDSHIRE, EAST WORCESTERSHIRE AND NORTH WARWICKSHIRE

As equipped with fineries and chaferies.

Forge	Location	River
Brewood (Lower)	Brewood	Penk
Brewood (Upper)	Brewood	Penk
Broadwaters (Lower)	Near Kidderminster	Broadwaters Brook
Bromwich Forge	West Bromwich	Tame
Bromford Mill	Erdington	Tame
Bromley	Abbots Bromley	Blithe
Burton	Burton-on-Trent	Trent
Cannock	Cannock	Rising Brook
Clatterbach	Stourbridge	Worcestershire Stour
Clifford	Stratford-upon-Avon	Warwickshire Stour
Corngreaves Forge	Cradley	Worcestershire Stour
Congreve	Penkridge	Penk
Cookley	Kidderminster	Worcestershire Stour
Coven	Coven	Penk
Cradley	Cradley	Worcestershire Stour
Deepmore	Near Coven	Wyrley Brook
Golds Hill	West Bromwich	Tame
Gothersley	Stourbridge	Smestow
Glasshampton	Astley	Dick Brook
Greens Forge	Stourbridge	Smestow
Halesowen Forge	Halesowen	Worcestershire Stour
Heath Mill	Womborne	Smestow
Hints	–	Bourne Brook
Hyde Mill	Kinver	Worcestershire Stour
Little Aston Forge	–	Bourne Brook
Lodge Forge	Cradley	Worcestershire Stour
Lower Mitton	Stourport	Worcestershire Stour
Lye Forge	Lye	Worcestershire Stour
Perry Barr Mill	Handsworth	Tame
Redditch	Redditch	Arrow
Rugeley	Rugeley	Rising Brook
Swindon	Swindon	Smestow
Toll End	Tipton	Tame

Troyal Forge	Stourbridge	Worcestershire Stour
Upper Mitton	Stourport	Worcestershire Stour
Wall Heath	Kingswinford	Holbeache Brook
Wednesbury	Wednesbury	Tame
Whittington	Kinver	Worcestershire Stour
Wilden	Stourport	Worcestershire Stour
Willoughby	Middleton	Tributary of Tame
Wolseley	–	Stafford Brook
Wolverley	Kidderminster	Worcestershire Stour
Wolverley (Lower)	Kidderminster	Worcestershire Stour

Lodge Forge, in the ownership of Coley and Townshend, was advertised for sale in 1852 as a water-powered forge.

LIST OF COKE BLAST FURNACES IN SOUTH STAFFORDSHIRE AND EAST WORCESTERSIRE, 1757–1900

1. Barborsfield
Also known as 'Ettingshall Furnaces'

Maximum no. of furnaces: 3
Established: 1825

Owners
1. Thomas Banks
2. Thomas Banks & Son
3. William Banks, trading as Thomas Banks & Son
4. Shale and Fowler
5. Barbors Field Co.

Canal access – BCN Old Main Line.
Banks & Son also operated the Ettingshall Ironworks on the opposite side of the canal.

2. Bentley

Maximum no. of furnaces: 4

Owners
1. Earl of Lichfield
2. Countess of Lichfield
3. Clarke and Jerome
4. William Riley
5. Chillington Iron Co.

Canal access – BCN Ansons Branch.
Siding access for Midland Railway at a subsequent date.
Private railway to pits, gauge 3ft 3in.

3. Bilston
(Renamed Spring Vale)

Maximum no. of furnaces: 5

Owners
1. William Bickley & Co.
2. William Sparrow
3. George Jones
4. John Jones of Staffordshire (and son to George Jones)
5. Jones and Murcot (renamed furnaces Spring Vale)
6. Alfred Hickman
7. Alfred Hickman Ltd

Beside private canal arm linked to BCN at Bilston.
Various tramway links.

a) Under Sparrow tenancy links principally north to mines Stow Heath, etc., and links with adjacent Bilston Ironworks.
b) Under Jones tenancy, links west to mines at Spring Vale Colliery and the Spring Vale Ironworks.

Subsequent transport links made with Alfred Hickman and successors.
Standard-gauge siding link with LNWR Stour Valley Railway (as Spring Vale) made c.1872.
Standard-gauge siding link with GWR (WMR) as Spring Vale.
The private railway links were extended to join with mills and Staffordshire Steel & Ingot Ironworks (1882–1884).

4. Bilston Brook

Maximum no. of furnaces: 3

Owners
1. Thomas Price
2. Thomas & Henry Price
3. Executors of T.B. Price
4. John Parsons
5. G.H. & A. Hickman
6. Samuel Griffiths
7. B. Gibbons junior
8. Emma Gibbons
9. Thomas and Isaac Bradley, Richard Dodd and John Southan, trading as Brook Furnaces Iron Co.
10. Thomas and Isaac Bradley and John Southan trading as Brook Furnaces Iron Co.

Located beside Bilston Branch (BCN Walsall Canal).
Narrow-gauge tramway links to mines.
For sale, 1881.

5. Birchills
Also known as:
a) 'Old Birchills'
b) 'Roughwood'

Maximum no. of furnaces: 2

Owners
1. Stubbs & James
2. S. Walker & Co.
3. E. Tyler
4. P. Williams
5. F.C. Perry
6. Williams Brothers (Henry and Philip Addison Williams)

Tramroad to private basin alongside the Wyrley & Essington Canal at Birchills.

6. Bovereaux
Also known as 'Lower Bovereaux'

Maximum no. of furnaces: 2
Established: *c.*1849

Owners
1. Baldwin & Son

2. Tame Iron Co. (T. Holcroft)

Beside BCN Old Main Line

7. Bradeshall
Also known as 'Stour Valley'

Maximum no. of furnaces: 2

Owners
1. B. Richards & Co.
2. W.J. & G. Onions (one furnace), Frederick Giles (one furnace)
3. James & George Onions
4. William & Edward Onions
5. John & Samuel Roberts

Beside BCN New Main Line.
Land leased from BCN in 1853 to erect furnaces.
J. & S. Roberts also owned the Swan and Small Heath Foundries, West Bromwich.

8. Bradley

Maximum no. of furnaces: 2

Owners
1. G.B. Thorneycroft & Co.

These furnaces were built at new location in 1860 to replace earlier furnaces at Upper Bradley. They were close to WMR with a tramway south to BCN Bradley Branch. G.B. Thorneycroft operated ironworks at Swan Garden and Shrubbery Ironworks, Wolverhampton.
Furnaces for sale, 1880.

9. Bradley Bridge

Maximum no. of furnaces: 2
Established: *c.*1866

Owners
1. David Jones
2. James Light
3. Bradley Bridge Charcoal Iron & Foundry Co. Ltd

Furnaces for sale, 1883.

10. Brettell Lane

Maximum no. of furnaces: 2
Established: 1825

Owners
1. Hills & Wheeley
2. Hall, Holcroft & Pearson

Furnaces and ironworks were located beside
the Stourbridge Canal.

11. Brierley Hill

Maximum no. of furnaces: 1

Owners
1. Izons & Co.

Furnace located on hillside at a distance from
Stourbridge Canal, but linked to it through
tramways.

12. Brierley Hill

Maximum no. of furnaces: 2
Established: 1796

Owners
1. Bancks & Onions
2. John Onions & Co.
3. John Bradley & Co.

Furnaces were located alongside the
   Stourbridge Canal.
There were also associated ironworks, which
   retained by John Bradley & Co.
The blast furnaces were demolished.

13. Broadwaters
Also known as 'Waterloo'

Maximum no. of furnaces: 3
Established: *c.*1815

Owners
1. W. Matthews
2. Horace St Paul, Thorneycroft & Co.
3. Colbourne, Groucutt & Co.
4. Samuel Groucutt
5. Samuel Groucutt & Sons

Located beside Walsall Canal at Broadwaters.
Tramways to collieries.
Samuel Groucutt & Sons also owned
   Bradleyfield Ironworks, Bilston and
   Bankfield Ironworks, Bilston.

14. Buffery
Renamed Dixon's Green

Maximum no. of furnaces: 1
Furnaces known as 'Dixons Green', renamed
'Buffery', *c.*1869

Owners
1. Joseph Haden
2. William Haden
3. Assignees of William Haden
4. John Jones & Sons

Beside Dudley No. 2 Canal

15. Buffery
Also known as 'New Buffery'

Maximum no. of furnaces: 3
Established: By 1812

Owners
1. Fereday, Jones & Co.
2. Blackwell, Jones & Co.
3. Molineaux & Co.

Out of blast by 1843.
Furnaces located at the Buffery Colliery and
   linked to Dudley Canal by a tramways.

16. Buffery
Also known as 'Old Buffery'

Maximum no. of furnaces: 1
Established: *c.*1813

Owners
1. Richard Salisbury & Co.

NG Tramways to mines.

17. Cape

Maximum no. of furnaces: 2

Owners
1. Cape Iron Co.
2. Benjamin Richards

Beside BCN Cape Arm at Smethwick.
One furnace was established by 1864.
  Subsequently two were recorded on site in 1868.
Both were out of blast by 1869.

18. Capponfield

Maximum no. of furnaces: 3
Established: By 1806

Owners
1. Smith, Read & Co.
2. W. Aston
3. J. Bagnall
4. Thomas and Isaac Bradley

Beside BCN Old Main Line.
Siding with GWR from *c.*1901.

19. Chillington

Maximum no. of furnaces: 4

Owners
1. Jones and Foster
2. Barker Brothers
3. Chillington Iron Co.
4. Chillington Iron Co. Ltd

Works were established in 1828, although parts
  of the mineral estates were at work earlier.
Canal Basin BCN Main Line.
Railway siding GWR via wagon turntable to
  Walsall Street Goods, Wolverhampton.
Private railway (3ft in gauge) linked canal,
  Chillington Ironworks and various mines at
  Moseley Hole and Portobello.

Other furnaces at Bentley (near Walsall and
Moseley Hole (Willenhall)).

Ironworks:
a) Chillington, Wolverhampton
b) Capponfield, Bilston
c) Leabrook, Wednesbury
d) Bradley, Bilston

Furnaces shut down 1884 and demolished
between 1885 and 1886.

20. Coltham
Also known as 'Essington'

Maximum no. of furnaces: 2

Owners
1. Mr Penn
2. John Tristham
3. Coltham Coal & Iron Co. (Daniel Bagnall, agent)
4. R. Mainwaring

Furnaces established by 1808 and placed beside
  Wyrley & Essington Canal, north of Willenhall,
  utilising initial supplies of ironstone and coal.
  By 1826 only one furnace is mentioned.
Furnaces infrequently in blast and completely
  disused by the mid-1840s.

21. Coneygre

Maximum no. of furnaces: 3
Established: 1794

Owners
1. Zachariah Parkes
2. George Parker
3. Earl of Dudley Trustees
4. Earl of Dudley

Beside BCN Old Main Line.
Standard-gauge siding to LNWR (SSR).
Furnaces dismantled and sold during 1896.

22. Corbyns Hall

Maximum no. of furnaces: 4
Established: 1825

Owners
1. John & Benjamin Gibbons
2. Matthews and Dudley
3. William Matthews & Co.
3a. One furnace to Samuel Hingley (1868–1871)
4. William Matthews & Co. Ltd

Connected by private tramway to Stourbridge
Canal.

Tramway replaced by Stourbridge Extension
Canal.

Sidings also made to connect with
Kingswinsford Branch Railway (OWWR)
after 1858.

Furnaces for sale, 1889.

23. Corbyns Hall
Also known as 'New Corbyns Hall'

Maximum no. of furnaces: 4
Opened: *c*.1840

Owners
1. B. Gibbons
2. Bromley Coal and Iron Co., Messr Parsons
   and Chambers – Manager: T.J. Parsons (other
   property includes Bromley Hall IW)

24. Corngreaves

Maximum no. of furnaces: 6

Owners
1. G. Attwood & Sons – Steelworks only
2. British Iron Co. (Small Shears and Taylor)
   – erected furnaces *c*.1826
3. New British Iron Co.
4. Corngreaves Iron Co.
5. R. Fellows

3ft 2½in-gauge tramway to collieries and other
furnaces at Dudley Wood.

NG Track later diverted to Fly Colliery Basin
(Dudley No. 2 Canal).

Standard-gauge sidings link to Stourbridge
Railway (WMR/GWR).

Congreaves Forge was originally a water-
powered mill that produced crucible steel and
steel goods.

Ironworks at Congreaves and Brierley Hill.

25. Coseley

Maximum no. of furnaces: 2

Owners
1. George Jones
2. Jos & Thomas Turley

Beside BCN Old Main Line.

26. Crookhay

Maximum no. of furnaces: 4

Owners
1. T. Davies & Sons
2. G. Thompson & Co.
3. W. &.G. Firmstone
4. H.O. Firmstone

Furnaces and mines at terminus of BCN
Dartmouth Branch.

W. & G. Firmstone operated also the Crookhay
Ironworks and Leys (Lays) Furnaces.

Furnaces for sale, 1888.

27. Darlaston Green

Maximum no. of furnaces: 3
First furnace erected by 1848

Owners
1. Bill and Mills
2. S. Mills
3. Darlaston Steel & Iron Co.
4. Darlaston Coal & Iron Co.
5. T. & I. Bradley

Beside BCN, Walsall Canal.

Standard-gauge siding link with LNWR
Wednesbury-Darlaston Branch provided for
Darlaston Steel & Iron Co.

One Furnace leased by Swan Iron Co. until
1880.

Thomas & Isaac Bradley operated two furnaces
from 1883.

28. Deepfields

Maximum no. of furnaces: 3
Established: *c*.1788

Owners
1. Stokes & Pemberton
2. Edwin Pemberton
3. Benton & Pemberton – (one furnace let to
   S. Groucutt & Sons)
4. Thomas Hook Pemberton
5. Deepfields Iron Co.

6. William Edgeworth Gibbons
7. Trustees Samuel Pemberton

Beside BCN Old Main Line.
Dismantled, 1892.

29. Deepfields

Maximum no. of furnaces: 2
Established: *c.*1812

1. Richard Hawkes
2. Benjamin Whitehouse & Co.

30. Dibdale

Maximum no. of furnaces: 1
Established: *c.*1802

Owners
1. Thomas Vernon
2. Fereday, Crockitt & Co.
3. John Crockitt, Edward Crockitt & Co.
4. Hornblower & Co.
5. Crockitt & Co.

Tramway to mines.

31. Dudley Port

Maximum no. of furnaces: 2
Established: 1824

Owners
1. J.T. Fereday
2. Horton & Gill
3. Hopkins & Son
4. William Hopkins

Basin beside BCN (Old Main Line).
Tramway access to mines and brickyard.
Furnaces for sale, 1868.

32. Dudley Port
Also known as 'Horseley Heath Furnace'

Maximum no. of furnaces: 1

Owners
1. James & George Onions

2. Edwin Onions

Beside BCN Dixon's Branch.

33. Dudley Wood
Also known as 'Netherton Four Furnaces'

Maximum no. of furnaces: 4
Established: *c.*1802

Owners
1. G. Attwood & Sons
2. British Iron Co.
3. New British Iron Co.
4. Noah Hingley

Beside Dudley No. 2 Canal.
NG tramway links to mines.

34. Eagle
Also known as 'Great Bridge'

Maximum no. of furnaces: 2
Established: *c.*1800

Owners
1. Richard Hawkes
2. Haynes & Hawkes
3. William Aston & William Baldwin
4. William Aston
5. Hills & Wheeley
6. J. Hartland

Located beside BCN, Walsall Canal.
Furnaces for sale in 1847 and demolished to make way for South Staffordshire Railway and Great Bridge Interchange basin.

35. Glebefield

Maximum no. of furnaces: 1

Owner
1. Thomas & Henry Price

36. Golds Green

Maximum no. of furnaces: 3
Established: 1820

Owners
1. John Bagnall & Sons
2. John Bagnall & Sons Ltd

The furnaces were placed beside BCN Balls Hill Branch.
Associated ironworks:
a) Golds Hill, West Bromwich
b) Lea Brook Ironworks, Tipton
c) Imperial Ironworks, Wednesbury
d) Toll End Ironworks

The original ironworks at Golds Hill was a water-powered mill beside River Tame, worked by various proprietors until acquired by the Bagnalls. Tramway links were made to Golds Hill Ironworks and mines with Goldsgreen Furnaces.
Furnaces for sale, 1882.

37. Gospel Oak

Maximum no. of furnaces: 4
Established: *c.*1792

Initially one furnace, a second furnace was erected by 1806 and another two more were added during 1828. Figures may include those erected at Tibbington (Tipton Old Church).

Owners
1. Read & Dumaresq
2. John Read
3. S. Walker & Co.

Beside BCN Old Main Line.
Link also by the private Dumaresq Branch to Gospel Oak Branch and Walsall Canal.
Furnaces taken down by 1840.
There was an adjacent ironworks which continued to be used after the furnaces were put out and demolished.

38. Graveyard

Maximum no. of furnaces: 1
Established: *c.*1798

Owners
1. Thomas Jones
2. Bagley & Co.

NG Tramway to mines.

39. Green Lane
Also known as:
a) 'Birchills'
b) 'Bloxwich'

Maximum no. of furnaces: 2

Owners
1. T. and C. Highway
2. J. Hartland
3. J. Jones & Son
4. G. & J. Jones
5. Walsall Iron Co.
6. Birchills Iron and Steel Co. Ltd

Basin link to Wyrley & Essington Canal.
SG siding link to SSR/LNWR Cannock Branch.

40. Groveland

Maximum no. of furnaces: 1
Established: *c.*1859

Owners
1. H. Richards
2. G.H. & A. Hickman
3. George Haden Hickman

Beside private canal arm to Groveland Colliery, link with BCN Old Main Line.
Also ironworks adjacent.

41. Hallen
Also known as 'Wednesbury Ironworks'

Maximum no. of furnaces: 1
Established: *c.*1785

Owners
1. John & Samuel Hallen
2. Assignees of J. & S. Hallen
3. John Wheeler

Furnace and ironworks were located beside Monway Branch (Walsall Canal).
John & Samuel Hallen were made bankrupt, but furnaces continued by assignees.
Ironworks advertised for sale, 1816.

42. Hallfields

Maximum no. of furnaces: 1
Established: *c.*1804

Owners
1. John Wilkinson
2. Fereday, Smith & Co.
3. J.T. Fereday
4. F.C. Perry
5. B. Gibbons junior
6. F.C. Perry
7. Moorcroft Colliery Co.

Beside BCN Old Main Line.

43. Hange
Also known as 'Tividale'

Maximum no. of furnaces: 2
Established: *c.*1867

Owners
1. Round Brothers

Beside BCN Old Main Line.

44. Hatherton

Maximum no. of furnaces: 2

Owners
1. Richard Fryer
2. William Fleeming Fryer
3. T. & C. Highway
4. G.B. Thorneycroft
5. W.F. Fryer
6. G. & R. Thomas

Beside Wyrley & Essington Canal.
2ft 10in-gauge tramway to mines.
Standard-gauge sidings to LNWR (SSR),
    locomotive haulage from *c.*1876.

45. Herberts Park
Also known as 'Darlaston'

Maximum no. of furnaces: 2

Owners
1. David Jones

Beside BCN Walsall Canal.
NG Tramways to mines.

46. Highfields

Maximum no. of furnaces: 2

Owners
1. J. & W. Firmstone

Erected on Highfields Colliery. Brief existence
    *c.*1820–1824.

47. Horseley

Maximum no. of furnaces: 2
Constructed between 1808 and 1809.

Owners
1. Horseley Iron Co.

Built beside BCN Toll End Communication
    canal and adjoining engineering works.

48. Horseley

Maximum no. of furnaces: 2
Constructed: *c.*1849

Owners
1. John Colbourn, Sons & Co.

Beside BCN Toll End Communication Canal.

49. Ketley

Maximum no. of furnaces: 3
Constructed, *c.*1840

Owners
1. Jones & Oaks until 1849
2. B. Gibbons

Furnaces for sale, 1861.

50. Lays

Maximum no. of furnaces: 3
Established: 1828

Owners

1. W. & G. Firmstone

Basin link to Stourbridge Canal.
Operated Crookhay Furnaces & Ironworks
and Oak Farm Furnaces.
D. & G. Horton and then Brown & Freer
operated adjacent Leys Ironworks.
Furnaces for sale, 1895.

51. Leabrook

Maximum no. of furnaces: 1
Established: *c.* 1810

Owners

1. Michael Tony
2. John Bagnall & Co.
3. Leabrook Furnace Co.
4. B. Caddick & Co.

Placed beside the turnpike road and near to
the Walsall Canal.

52. Lower Bradley

Maximum no. of furnaces: 1
Established: 1758

Owners

1. John Wilkinson
2. J. Cooper

Furnace was located near road at Fiery Holes.
There was also an ironworks and rolling mill
worked by Hallen & Cooper until 1827.

53. Millfields

Maximum no. of furnaces: 3

Owners

1. Fereday, Walker & Co.
2. J. Walker
3. W. Riley
4. B. Gibbons
5. Mrs E. Gibbons
6. W. & J.S. Sparrow

Beside BCN Main Line near Ettingshall.
Tramway north to mines.

Later standard-gauge siding link with GWR/
OWWR.

54. Moorcroft

Maximum no. of furnaces: 2
Established: *c.* 1802

Owners

1. Edward, Henry & John Addenbrooke

Furnaces placed beside private Bradley Branch
Canal.
Disused by 1840.

55. Moseley Hole

Maximum no. of furnaces: 3

Owners

1. Chillington Iron Co.

3ft-gauge tramway – horse and later
locomotive-worked from Chillington Works
and to mines at Portobello.
Standard-gauge sidings with GJR/LNWR,
also locomotive-worked.

56. Moxley
Also known as 'Prince of Wales'

Maximum no. of furnaces: 2
First furnace erected: 1871

Owners

1. David Rose

Beside Walsall Canal.
Standard-gauge railway made between
furnaces and LNWR at Wednesbury worked
by their own locomotive.
Worked in association with adjacent
ironworks, Albert, Victoria and Moxley
Forge.
Furnaces for sale, 1892.

57. Netherton
Also known as 'Blowers Green'

Maximum no. of furnaces: 4

Owners
1. M. & W. Grazebrook

Beside Dudley No. 2 (or Netherton) Canal.
Narrow-gauge tramways to mines.
Later standard-gauge siding to WMR/GWR,
    locomotive-worked.

58. Netherton
Also known as:
a) 'Netherton Two Furnaces'
b) 'Bumble Hole Furnaces'

Maximum no. of furnaces: 2
Established: *c.*1814

Owners
1. Christopher Bancks
2. G. Attwood & Sons
3. British Iron Co.
4. New British Iron Co.
5. N. Hingley & Sons
6. James & George Onions
7. Joseph. H. Pearson
8. Baldwins Ltd

Beside Dudley No. 2 Canal.
Later standard-gauge siding link to GWR
    Wythymoor Basin Branch.

59. New Bilston
Previously Bilston

Maximum no. of furnaces: 5

Owners
1. W. Baldwin & Co. (Bilston)
2. Blackwell & Co. (New Bilston)

Beside BCN Bilston Branch.
Disused by 1861 and demolished, *c.*1867.

60. New Birchills
Renamed – Castle

Maximum no. of furnaces: 5

Owners
1. George Jones
2. John Jones
3. J. Brayford

4. Birchills Estates Co.
5. Castle Coal and Iron Co.
6. Castle Iron Co.

Beside Wyrley & Essington Canal.
Standard-gauge siding links with LNWR/SSR.

61. New Level

Maximum no. of furnaces: 5
Established: *c.*1801

Owners
1. Benjamin Gibbons
2. Earl of Dudley's Trustees (from 1844)
3. Earl of Dudley
4. Round Oak Steelworks Ltd (from 1891)

Beside Dudley Canal.
Rebuilt and reconstructed 1844–1845.
Served by 2ft 6in-gauge tramway systems
    and standard-gauge internal system – both
    locomotive-worked.
Also Coneygree Furnaces, Tipton and Round
    Oak Works, Brierley Hill.

62. New Priestfield

Maximum no. of furnaces: 2
Established: 1859

Owners
1. William Ward & Sons
2. Patent Shaft & Axletree Co.

Beside BCN Walsall Canal.
NG Tramway from LNWR at Willenhall
    – locomotive-worked.
Line later regauged to standard.

63. Oak Farm

Maximum no. of furnaces: 2
Comprised furnaces, ironworks and foundry.

Owners
1. Oak Farm Coal & Iron Co. (Patterson,
    Walker, Boydell & Roper)
2. S.H. Blackwell and Henry Onions
    Firmstone trading as Oak Farm Furnace Co.
3. Henry Onions Firmstone

Company first formed in 1835. Following the
break up of the Oak Farm Coal and Iron
Co. (1848), Messr Warden and Williams
leased the ironworks and Emanuel Cookson
leased the foundry.

All works were located near end of
Stourbridge Extension Canal.

64. Old Hill

Maximum no. of furnaces: 2
Established: 1848

Owners
1. Thomas & Isaac Badger
2. David Rose
3. Noah Hingley & Sons

Beside Dudley No. 2 Canal.
Badger also owned an ironworks at Dudley
Port that was formerly worked by Zachariah
Parkes & Co.
Hingleys worked furnaces in association with
Netherton Ironworks.

65. Old Level

Maximum no. of furnaces: 2
Established: 1786
Originally known as the 'Level Furnaces'.

Owners
1. Richard Croft
2. W. Izons & Co.
3. Hall, Holcroft & Pearson
4. James Holcroft

Demolished: 1897

66. Old Park, Wednesbury

Maximum no. of furnaces: 3

Owners
1. Lloyds Foster & Co.
2. Patent Shaft and Axletree Co. Ltd

Tramway to Walsall Canal, horse-worked, and
later locomotives used for haulage.
Internal railway system served Monway Works.
Furnaces dismantled during 1877.

67. Oldbury

Maximum no. of furnaces: 4

Owners
1. George Parker & Co.
2. John Dawes & Sons
3. William Bennitt (Captain Bennitt)
4. Edward Onions & Co.
5. J. & S. Onions

Tramway to mines.
Beside BCN Old Main Line.
Dismantled after 1872.

68. Osier Bed

Maximum no. of furnaces: 3

Owners
1. Osier Bed Iron Co. (William Hanbury
Sparrow)

Narrow-gauge tramway link to Stow Heath
Furnaces and also LNWR.
Also owned the Horsley Fields iron and
tinplate works, Wolverhampton.
Furnaces dismantled during 1885.

69. Park Lane

Maximum no. of furnaces: 2

Owners
1. Thomas Morris
2. Thomas Morris Jr
3. John Colbourne, Sons & Co.

Beside BCN New Main Line.
For sale, 1884.

70. Parkfield

Maximum no. of furnaces: 5
Established: 1825

Owners
1. Bishton & Co.
2. Parkfield Iron Co. (Bishton & Co.)
3. Dimmack & Marten
4. Parkfield Iron Co. Ltd

5. J. & T. Williams

Tramway to BCN, horse and locomotive-
  worked.
Tramway to LNWR at Ettingshall.
Dismantled during 1888.

71. Parkhead

Maximum no. of furnaces: 2
Established: 1800

Owners
1. Zachariah Parkes & Co.
2. Evers & Martin
3. Phillips & McEwen

Beside Dudley Canal.

72. Pelsall

Maximum no. of furnaces: 2

Owners
1. Richard Fryer
2. W.F. Fryer
3. Davies & Bloomer
4. Boaz Bloomer & Son
5. Pelsall Coal Iron Co. Ltd

Beside Wyrley & Essington Canal.
Standard-gauge railway – Locomotive-worked
  from 1865.
Boaz Bloomer & Son also owners of Team
  Valley Works, Gateshead, Co. Durham.

73. Priestfield

Maximum no. of furnaces: 3
Established: *c.*1808

Owners
1. Fereday, Turton & Co.
2. William Ward
3. William Ward & Sons

Tramway link to Birmingham Canal at Millfields.

74. Priorsfield
Also known as 'Deepfield'

Maximum no. of furnaces: 3

Owners
1. Henry Bickerton Whitehouse
2. H.B. Whitehouse & Son

Served by Birmingham Canal.
Later siding made from Furnaces to London
  & North Western Railway and worked
  by LNWR engines and a steam crane at
  furnaces.

75. Ridgacre

Maximum no. of furnaces: 1

Owners
1. P. Fowler

Out of blast by 1850

76. Rough Hay

Maximum no. of furnaces: 3
Established: 1843

Owners
1. Addenbrooke, Smith and Pidcock

Beside BCN Walsall Canal.
NG Railway (probably 3ft 3in gauge) to mines.
Dismantled during 1883, part of plant
  transferred to Wellingborough Furnaces,
  Northamptonshire.

77. Roughills

Maximum number of furnaces: 3

Owners
1. Fereday & Co.
2. John & William Firmstone
3. William Aston & Co.

Basin link with Birmingham Canal and
  tramways to furnaces.

78. Russells Hall

Maximum no. of furnaces: 5
Established: 1827

Owners
1. Morrison & Co.
2. D. & G. Horton
3. Blackwell & Co.
4. S.H. Blackwell

Tramway to mines, Dudley Canal and Pensnett Canal.
Disused by 1861.

79. Stonefield

Maximum no. of furnaces: 1

Owners
1. Edward Woolley
2. G.H. & A. Hickman
3. Stonefield Iron Co.
4. Bilston Iron Co.

Some blast furnace statistics mention Thomas Crew as a proprietor of Stonefield. Some entries contains errors and more research is needed to link Crew (of the Buffery Collieries) with this Bilston Furnace.

80. Stow Heath

Maximum no. of furnaces: 5
Established: 1824

Owners
1. William Hanbury Sparrow and John Sly Sparrow
2. W. & J.S. Sparrow

Tramway to BCN and Bilston Ironworks.
Disused by 1879.

81. Shut End

Maximum no. of furnaces: 4
Established: *c.*1829

Owners
1. John Bradley & Co.

Standard-gauge railway link to Staffordshire & Worcestershire Canal at Ashwood Basin completed 1829. Railway originally comprised a central level worked by the

steam locomotive *Agenoria*.
Served by Stourbridge Extension Canal from ironworks at Shut End, Brockmoor and Brierley Hill.

82. Tibbington
Also known as 'Old Church'

Maximum no. of furnaces: 2
Shown on plan of estate, 1817

Owners:
1. John Read
2. S. Walker & Co.
3. John & Edmund Walker

There was also an ironworks plant known as 'Old Church'.
Works were located beside private basin that joined Old Main Line near Tibbington church.
Furnaces disused by 1840.

83. Tipton
Also known as:
a) 'Tipton Green'
b) 'Parker's Old Furnace'

Maximum no. of furnaces: 2
Established: *c.*1781

Owners
1. George Parker & Co.

Near Junction with Lord Ward's Arm – Later Dudley Canal.
There was an ironworks adjacent to furnaces.

84. Tipton
Also known as 'Parker's New Furnace'
Renamed 'Coseley Moor'

Maximum no. of furnaces: 2

Owners
1. George Parker & Co.
2. Edward Cresswell
3. Rhos Hall Iron Co.
4. Thomas Turley
5. Wones Bros

Beside BCN Old Main Line.

85. Tipton Green

Maximum no. of furnaces: 4
Established: 1808

Owners
1. Fereday & Co.
2. Tipton Green Furnace Co.
3. Gibbons & Roberts
4. Roberts & Co.

William & Alfred Roberts partners.
At terminus of BCN Three Furnace Branch.
At end of private branch from Toll End
   Communication.
Basin link with Old Main Line.
Standard-gauge siding connection with
   LNWR.
Locomotive haulage in sidings from *c.*1890.

86. Toll End

Maximum no. of furnaces: 2

Owners
1. Birmingham Coal Co.
2. Motteram and Deeley
3. Toll End Co.

Dismantling sale, 1862.

87. Upper Bradley

Maximum no. of furnaces: 2

Owners
1. John Wilkinson
2. John T. Fereday
3. G.B. Thorneycroft & Co.

88. Union

Maximum no. of furnaces: 3

Owners
1. Philip Williams & Sons
2. Stour Valley Coal & Iron Co. Ltd

Beside BCN New Main Line.

Standard-gauge sidings links with LNWR.
For sale, 1884.

89. Nallbrook

Maximum no. of furnaces: 2

Owners
1. Whitehouse & Grazebrook
2. Benjamin Whitehouse
3. H.B. Whitehouse

Beside BCN Bloomfield – Deepfield Canal
   link.

90. Wednesbury Oak

Maximum no. of furnaces: 3

Owners
1. Gibbons, Williams and Whitehouse
2. P. Williams & Sons

Beside BCN Old Main Line.
Standard-gauge sidings joined LNWR
   Princess End Branch.

91. Willenhall

Maximum no. of furnaces: 3
Established: 1854

Owners
1. Fletcher, Solly & Urwick
2. Willenhall Furnaces Ltd

2ft 6in-gauge tramway locomotive hauled.
Standard-gauge siding with Wolverhampton
   and Walsall Railway/MR.
Furnaces for sale, 1882.

92. Willingsworth

Maximum no. of furnaces: 3

Owners
1. Sir Horace St Paul
2. Haines & Co.
3. Job Haines & Son
4. Willingsworth Iron Co.

New Level
Furnaces,
Brierley Hill.

Beside Gospel Oak Branch Canal.

93. Windmill End

Maximum no. of furnaces: 3

Owners
1. Christopher Bancks, Lathum Blacker, William Benbow
2. Banks & Blacker
3. Christopher & William Bancks
4. Jones Fereday & Co.
5. Sir Horace St Paul
6. S. Partridge
7. Woodall & Smith
8. George Pell
9. Samuel Griffiths
10. Hickman & Co.
11. J. & G. Onions
12. J.H. Pearson

Beside Dudley No. 2 Canal.

94. Withymoor

Maximum no. of furnaces: 2
Established: *c.* 1840

Owners
1. Wythymoor Furnace Co.

2. Best & Barrs
3. W.H. Dawes
4. John and George Dunn

Near terminus of BCN Withymoor Branch.
For sale, 1880.

95. Wolverhampton

Maximum no. of furnaces: 3
Established: 1825

Owners
1. Timmins & Co.
2. Dixon, Thorneycroft & Head
3. Dixon, Neve & Co.
4. Edward Poole & Co.
5. Aston and Corns
6. Isaiah Aston & Co.

Out of blast by 1868.

96. Woodside

Maximum no. of furnaces: 3

Owners
1. Cochrane & Co.

Beside Dudley No. 1 Canal.

APPENDIX FOUR

STEAM-POWERED FORGES AND ROLLING MILLS IN SOUTH STAFFORDSHIRE, NORTH WARWICKSHIRE AND EAST WORCESTERSHIRE, 1790–1844

The following list details works, which were known to have employed puddling furnaces from the development of the Cort process (1783).

★ Previously a water-powered mill

No.	Ironworks	Location	Owners
1	Albion	West Bromwich	Walter Williams
2	Aqueduct	Birmingham	Hooton, Richards & Wilkes
3	Bankfield	Bilston	(1) Joseph Maybury & Co. (2) G. Jellicoe (3) W. Baldwin
4	Bilston (Bilston Mill)	Bilston	(1) Bickley & Co. (2) W. & J.S. Sparrow
5	Bloomfield	Tipton	Bradley, Barrows & Hall
6	Bordesley	Birmingham	Daniel Moore
7	Brades	Oldbury	William Hunt & Sons
8	Bradley (New Bradley)	Bilston	(1) James Turley (2) John Bradley & Co.
9	Bradley (Upper Bradley)	Bilston	(1) John Wilkinson (2) Fereday, Smith & Co. (3) John Turton Fereday (4) Edward & George Thorneycroft (5) G.B. Thorneycroft & Co.
10	Bovereaux	Bilston	William Baldwin & Co.
11	Brettell Lane	Brierley Hill	John Wheeley & Co.

No.	Ironworks	Location	Owners
12	Brierley Hill	Brierley Hill	(1) John Onions & Son (2) J. Bradley & Co.
13	Broadwaters★	Kidderminster	Addenbrooke & Co.
14	Brockmore Forge	Pensnett	Addenbrooke & Co.
15	BrockmoreForge	Pensnett	Emus, Sanders & Heywood
16	Bromford	West Bromwich	(1) John & Samuel Dawes (2) John Dawes & Sons
17	Capponfield	Bilston	(1) John Read (2) Banks & Ottway (3) Parkes & Ottway (4) Bishton & Underhill (5) John Bradley & Co.
18	Chillington	Wolverhampton	Chillington Iron Co.
19	Corngreaves	Cradley	British Iron Co.
20	Cradley Forge★	Cradley	Cradley Iron Co.
21	Darlaston Green	Darlaston	(1) Bills & Mills (2) Samuel Mills
22	Deepfields	Coseley	(1) William Aston (2) Charles Lloyd Browning
23	District (Smethwick)	Smethwick	Premises converted from Smethwick Brassworks during 1834 (1) Joseph Hadley
24	Dudley Port	Tipton	(1) Zachariah Parkes (2) Thomas & Isaac Badger
25	Dudley Port	Tipton	(1) S. Plant & Sons (2) Plant & Fisher
26	Eagle (Bromwich Forge)	West Bromwich	(1) T.B. & H. Price (2) Eagle Coal & Iron Co.
27	Ettingshall	Bilston	(1) Thomas Banks (2) Thomas Banks & Sons
28	Factory (Factory Forge) (Tipton Forge)	Tipton	(1) J.T. Fereday, J.P. Firmstone & W. Turton

No.	Ironworks	Location	Owners
			(2) John Turton Fereday
			(3) Richard Bradley agent for Sir Horace St Paul
29	French Walls	Smethwick	Bordesley Iron & Steel Co. Ltd
30	Golds Green	West Bromwich	(1) John Bagnall
			(2) John Bagnall & Sons
31	Golds Hill★	West Bromwich	(1) John Read
			(2) Thomas Hill
			(3) William Aston
			(4) John Bagnall & Sons
32	Gospel Oak	Tipton	(1) John Read
			(2) Samuel Walker & Co.
			(3) John & Edmund Walker
33	Great Bridge	West Bromwich	Thomas Stokes For sale, 1824
34	Great Bridge	West Bromwich	James Walton For sale, 1840
35	Greets Green	West Bromwich	(1) George Allarton & Josiah Wynn
			(2) Josiah Wynn trading Allarton & Wynn (or Winn)
36	Halesowen Forge	Halesowen	John Burr For sale, 1825 & 1827
37	Heath★	Wombourn	(1) Richard Jevons
			(2) Daniel Moore
			(3) Thomas Homfray
			(4) Gibbin, Smith & Goode For sale, 1826
38	Highfields	Bilston	(1) W. Parsons & Co.
			(2) W. & J.P. Firmstone
			(3) W. Riley & Son
39	Horseley	Tipton	(1) York, Harrison & Co. trading as Horseley Iron Co.
			(2) Bramah, Cochrane & Deeley
40	Hyde★	Kinver	(1) Francis Homfray
			(2) Thomas Homfray
			(3) Smith & Shepherd
			(4) Thomas Morris
			(5) Lee and Bolton
41	Imperial	Wednesbury	Russell Brothers For sale, 1844

No.	Ironworks	Location	Owners
42	Lays	Brierley Hill	(1) Ann Horton & Co (2) Daniel & George Horton (3) Brown and Freer
43	Leabrook (Leabrook Forge)	Tipton	(1) Michael Tony (2) Thomas Bolton & Co. (3) Devey, Sanders & Tickell (3) Tarrett, Timmins & Co. (4) David & Joseph Smith (5) James Turley
44	Lower Bradley	Bilston	(1) John Wilkinson (2) Fereday Smith & Co. (3) John Turton Fereday (4) Turley, Dunning & Co. (5) Hallen, Cooper & Homer For sale, 1827
45	Lye Forge★	Stourbridge	John Folkes
46	Millfields	Bilston	(1) John Walker (2) Emily Walker (3) William Riley
47	Monway Field	Wedsnesbury	Lowe & Field For sale, 1839
48	New Smethwick	Smethwick	(1) Wilson Lloyd (2) Bordesley Steel & Iron Co. (3) Spittle & Downing
49	Nine Locks (Brierley Hill)	Brierley Hill	(1) Hornblower & Smith (2) Serjeant & William Hornblower (3) British Iron Co.
50	Old Church (Tibbington)	Tipton	(1) John Read (2) Samuel Walker & Co. (3) John & Edmund Walker
51	Old Church	Tipton	(1) Round & Caddick (2) John Henderson
52	Old Park	Wednesbury	Lloyds, Foster & Co.
53	Pelsall	Pelsall	(1) Richard Fryer (2) W.F. Fryer
54	Roway	West Bromwich	Edward Page & Sons

No.	Ironworks	Location	Owners
55	Shrubbery	Wolverhampton	George & Edward Thorneycroft
56	Stambermill Forge★	Stourbridge	John Fellows
57	Stourbridge	Stourbridge	J. Bradley & Co.
58	Summerhill	Tipton	(1) William & Isaiah Millington
59	Swindon★	Swindon	(1) Edward & George Benjamin Thorneycroft
60	Tipton	Tipton	(1) George Parker & Co. (2) Edward Creswell & Sons
61	Tipton Green (Upper Green Forge)	Tipton	(1) Thomas Hill (2) Mary Hill & Son (3) Mary Hill (4) Joshua Hill
62	Tividale	Tipton	(1) Thomas & Richard Homer (2) Hollis Solly & Sons
63	Toll End★	Tipton	Daniel Moore
64	Toll End	Tipton	(1) Wooley and Pretty (2) Kempson, Grafton & Brinton (3) Kempson & Grafton
65	Wednesbury	Wednesbury	(1) J. & G. Hallen (2) John Wheeler For sale, 1817
66	Wednesbury Forge★	Wednesbury	(1) John Wood (2) Joseph Round
67	Wednesbury Oak	Tipton	(1) Phillip Williams (2) Phillip Williams & Sons
68	Whittington★	Kinver	(1) Hartshorne & Podmore (2) Warden & Williams
69	Wilden★	Stourport	(1) Pratt & Co. (2) Lewty & Co.
70	Willenhall Forge	Willenhall	For sale, 1816
71	Wolverley★	Kidderminster	David Humphreys

APPENDIX FIVE

STEAM-POWERED FORGES AND ROLLING MILLS IN SOUTH STAFFORDSHIRE, NORTH WARWICKSHIRE AND EAST WORCESTERSHIRE, 1845–1900

The following list details of works which were known to have employed puddling furnaces, from the development of the Hall process (1839) until 1900.

Puddling furnace statistics compiled from mineral statistics and sales of plant.

PF Maximum number of puddling furnaces.

$ Number of puddling furnaces not known.

★ Previously a water-powered mill.

No.	Ironworks	Location	Owners	PF
1	Albert	Darlaston	(1) Fletcher, Rose & Co.	21
			(2) David Rose	20
			(3) Moxley Sheet Iron Co.	
2	Albert	West Bromwich	(1) James Shenton	8
			(2) Albert Sheet Iron Co.	
3	Albion	West Bromwich	(1) Walter Williams' Works split into Albion & Britannia 1860 on retirement of Mr Williams	50
4	Albion	West Bromwich	Albion Sheet Iron Co.	10
5	Alpha	Tipton	Joshua Wilkinson & Sons	8
6	Anchor	Smethwick	(1) William Cox	
			(2) Boydell & Glazier	
			(3) James Batson	
			(4) James Batson and Frederick William Ratcliffe	7
			(5) F.W. Ratcliffe	6
7	Atlas Gun Ironworks	Birmingham	William Bissell & John Millward Talbot trading as Bissell, Talbot & Co.	$
8	Aston Junction	Aston	Becks & Brookes	5
			A.J. Becks	
9	Atlas	West Bromwich	(1) Atlas Iron Co.	13

No.	Ironworks	Location	Owners	PF
			(2) Israel Parks	15
			(3) Parks and Parks	10
10	Bankfield	Bilston	(1) W. Baldwin	
			(2) Coulborne & Groucutt	
			(3) S. Groucutt & Sons	22
11	Batmans Hill	Bilston	(1) William Rose	12
			(2) Tupper & Co.	
12	Bescot Forge	Darlaston	(1) F.H. Lloyd	13
			(2) Bescot Forge Co.	
13	Bilston (Bilston Mill)	Bilston	W. & J.S. Sparrow	32
14	Birchills	Walsall	(1) John Brayford & John Lancaster	12
			(2) Bissell, Keay & Bissell	12
			(3) James Bissell & Son	13
			(4) George & Richard Thomas	13
15	Birchills Hall (Bloxwich)	Walsall	(1) Birchills Estates	
			(2) Birchills Hall Iron Co. Ltd	12
			(3) Bloxwich Iron & Steel Co. Ltd	12
16	Bloomfield	Tipton	(1) Bradley, Barrows & Hall	
			(2) Bramah, Barrows & Hall	
			(3) Barrows & Hall	56
			(4) William Barrows & Son	57
17	Brades	Oldbury	William Hunt & Sons	12
18	Bradford (Greets Green)	West Bromwich	Bradford Iron Co.	10
19	Bradley (New Bradley)	Bilston	(1) James Turley	
			(2) John Bradley & Co.	
			(3) Chillington Iron Co.	14
20	Bradley (Upper Bradley)	Bilston	G.B. Thorneycroft & Co.	
			For sale, 1854 and subsequently demolished	
21	Bradley (New Bradley)	Bilston	(1) Rose, Higgins & Rose,	9
			(2) Thomas Rose	
			(3) Gittings & Tolley	11
			(4) John Gittings	
			For sale, 1880	

No.	Ironworks	Location	Owners	PF
22	Bradleyfield	Bilston	(1) Rose, Higgins & Rose (2) Samuel Groucutt & Sons (3) World Galvanised Iron Co.	33
23	Bradley Hall	Bilston	(1) Burford, Thompson & Hadley (2) Thomas & Thompson (3) J.E. East & Co. For sale, 1896	10 10 10
24	Bradley Tinplate (Bovereaux)	Bilston	(1) William Baldwin & Co. (2) Thomson Hatton & Co. (3) Hatton Sons & Co. (4) Hattons Ltd (5) Joseph Sankey & Sons	6 6 11
25	Brettell Lane	Brierley Hill	(1) John Wheeley & Co. (2) South Staffordshire Ironworks Co. Ltd (2) Brettell Lane Iron Co. (3) Roberts and Cooper	16 16
26	Bretwell Hall	Brierley Hill	(1) T. Webb & Sons (2) South Staffordshire Ironworks Co. Ltd	6
27	Brickhouse	West Bromwich	Robert Williams & Son	8
28	Bridge (Bilston Brook)	Bilston	David Jones	8
29	Brierley Hill	Brierley Hill	J. Bradley & Co.	38
30	Britannia	Bilston	(1) Brereton, Holland & Cole (2) John Harris	9 11
31	Britannia	Oldbury	Bright, Perry, Gettings	13
32	Britannia	West Bromwich	(1) Samuel Griffiths & E.B. Thorneycroft (2) Britannia Iron Co. (3) Tupper & Co.	10 10
33	Broadwaters★	Kidderminster	(1) Banks & Morton (2) Thompson, Hatton & Co.	5 11
34	Brockmoor Tinplate	Pensnett	(1) Smith & Summerhill (2) Budd & Co.	18
35	Brockmoor	Pensnett	John Bradley & Co.	$
36	Bromley	Pensnett	(1) John Raybould (2) Roberts & Cooper	16

No.	Ironworks	Location	Owners	PF
37	Bromford	West Bromwich	(1) John Dawes & Sons	69
			(2) Bromford Iron Co.	65
38	Brunswick	Wednesbury	(1) Hardy, Hodgetts & Gibson	
			(2) Charles Geach & Co.	
			(2) J. Walker & Co.	48
			(3) Patent Shaft & Axletree Co.	54
39	Bulls Bridge	Darlaston	(1) Daniel Rose	10
			(2) David Skidmore	10
			(3) William Molineaux & Co.	12
			(4) Thomas Jevons	10
40	Bush Farm	West Bromwich	John J. Bowater & Sons	14
41	Cape	Smethwick	Benjamin Richards	7
42	Cape Town (Junction)	Smethwick	(1) Brookes & Beck	6
			(2) William Richard Brookes	6
43	Capponfield	Bilston	(1) Chillington Iron Co.	16
			(2) William Molineaux & Co.	20
44	Chillington	Wolverhampton	Chillington Iron Co.	66
45	Church Lane	Tipton	George Gadd& Co.	14
46	Church Lane (Bedford Street)	Tipton	(1) Joseph and George Onions	10
			(2) Waring & W. Onions	13
			(3) J. Jones	13
			(4) Hoddell & Whitehouse	13
47	Cleveland	Wolverhampton	(1) Wright & North	9
			(2) John & Isaac James Jenks trading as Cleveland Iron Co.	10
48	Cookley★	Kidderminste	(1) Cookley Iron Co.	12
			(2) J. Knight & Co.	18
			Works closed production removed to Brockamoor 1887 where iron replaced with steel	
49	Corbyns Hall	Kingswinsford	(1) Galvanised Iron Co.	
			(2) William Matthews	
			(3) Henry Sparrow	40
			(4) Corbyns Hall Iron Co.	40
			(5) Pensnett Iron Co.	

No.	Ironworks	Location	Owners	PF
50	Corngreaves	Cradley	(1) British Iron Co.	
			(2) New British Iron Co.	59
			(3) New British Iron Co. Ltd	
51	Cradley Forge★	Cradley	(1) Cradley Iron Co.	
			(2) Evers & Williams	
			(3) Samuel Evers & Sons	17
52	Crookhay	West Bromwich	(1) Thomas Davies & Sons	
			(2) George Thompson & Co.	18
			(3) Thompson & Firmstone	18
			(4) W. & G. Firmstone	18
			(5) Chambers, Chance & Chambers	17
			(6) Crookhay Iron and Steel Co.	16
53	Crown (Smethwick New Works)	Smethwick	(1) Downing & Spittle	
			(2) Edward Page	
			(3) J. Nicklin	10
			(4) Thomas L. Nicklin & Co.	11
			For sale, 1895	
54	Cyclops	Walsall	(1) Edward Russell	16
			(2) Exors of late Edward Russell	14
			(3) John Russell & Co. Ltd	21
55	Darlaston Green	Darlaston	(1) Bills & Mills	
			(2) Samuel Mills	22
			(3) Darlaston Steel & Iron Co.	22
			(4) Tolley, Sons & Bostock	14
56	Deepfields Sheet Ironworks (Deepfields)	Coseley	Owned by Pemberton family, various occupiers	
			(1) Browning & Jackson	11
			(2) Gibbs Brothers	10
			(3) E.J. Gibbs	
			(4) George Tinn & Co.	20
			(5) Coseley and District Iron Co. Ltd	
57	Dial	Amblecote	John Onions trading as British & Foreign Charcoal Iron Co.	$
58	District	Smethwick	(1) Beasley & Farmer	20
			(2) Patent Nut and Bolt	12
			(3) District Iron and Steel Co.	20
59	Dixons Green	Dudley	(1) William Haden	
			(2) Dixons Green Iron Co.	11
			(3) Frederick Cresswell	

No.	Ironworks	Location	Owners	PF
60	Dudley Port (Portfield)	Tipton	(1) Thomas & Isaac Badger (2) F. Giles (3) James Holcroft (4) James and Charles Holcroft	14 14 16
61	Dudley Port	Tipton	(1) S. Plant & Sons (2) Plant & Fisher For sale, 1898	20
62	Dudley Port	Tipton	Joseph Welch & Sons	8
63	Dunkirk (Dunkirk Forge)	West Bromwich	(1) Palmer & Spencer (2) T. Spencer (3) Thomas Jordan & Co.	9
64	Eagle	Oldbury	F. Simpson & Co.	19
65	Eagle	West Bromwich	Known as 'Bromwich Forge' when owned by T.B. Price (1) Eagle Coal & Iron Co.	14
66	Ebeneezer	Oldbury	(1) J.W. Adlington (2) Bright, Gettings & Perry (3) Thomas Perry For sale, 1881	5
67	Ebeneezer	Coseley	(1) Joseph Welch & Sons (2) H. Onions & Co. (2) James & William Sanders (3) E. Birch & Sons	
68	Ettingshall	Bilston	(1) Thomas Banks & Son (2) William Banks (3) Morewood & Co. Idle for ten years, 1876–1886 (4) Ettingshall Iron Co. (5) Davies Brothers & Co. For Sale, 1900	12 10
69	Excelsior	West Bromwich	Thomas P. Allen & Co. For Sale, 1886	12
70	Eyre Street (Spring Hill Sheet Mills)	Birmingham	(1) J. Fletcher & Co. (2) Beard & Eberhard	4

No.	Ironworks	Location	Owners	PF
71	Factory (Factory Forge) (Tipton Forge)	Tipton	(1) Richard Bradley & Son (2) Barrows & Hall (3) W. Barrows & Son (4) John Stevens & Co.	22
72	Globe	Tipton	(1) Tividale Iron Co. (2) Whitehead & Haynes (3) J.O. Haynes	9 / 4
73	Golds Green	West Browich	(1) John Bagnall & Sons (2) John Bagnall & Sons Ltd	$
74	Golds Hill★	West Bromwich	(1) John Bagnall & Sons (2) John Bagnall & Sons Ltd	31
75	Gospel Oak	Tipton	John & Edmund Walker	$
76	Gothersley★	Stourbridge	(1) Wm Fennemore (2) Hatton & Bunn	$
77	Great Bridge	West Bromwich	Lees & Holden / Hipkiss & Co.	10 / 5
78	Great Bridge	Tipton	(1) Walter Williams (2) E.B. Thorneycroft & Co.	12 / 12
79	Great Bridge	Tipton	(1) Solly Brothers (2) Great Bridge Iron and Steel Co. Ltd	18 / 24
80	Greets Green (Staffordshire)	West Bromwich	(1) Thomas Payne (2) Brayford & Lancaster (3) Croft & Son (4) E.B. Thorneycroft & Samuel Griffith (5) Thomas Cross (6) Eagle Coal & Iron Co. (7) William Jeffries (8) Star Iron and Tinplate Co. (9) Greets Green Iron and Steel Sheet	20 / 17 / 13 / 14 / 8 / 8
81	Grove (Phoenix)	Smethwick	(1) Nash, Hayward & Co. (2) Stephen Nash & George Hayward	10 / 10
82	Grove Street	Smethwick	(1) John Stones & Son (John & George Stones) (2) Joseph Stones & Co.	10 / 3
83	Groveland	Tipton	(1) G.H. & A. Hickman, (2) George Haden Hickman	18 / 24

No.	Ironworks	Location	Owners	PF
84	Gun Barrel	Smethwick	William Marshall	4
85	Church Lane	West Bromwich	Thomas Johnson Jr	7
86	Hall End (Church Lane)	West Bromwich	(1) Johnson & Co., (2) Thomas Johnson Jr (3) J.T. & W. E. Johnson	7 7 9
87	Harts Hill	Brierley Hill	(1) William Jeffries (2) Hingley and Smith	30 36
88	Herberts Park	Darlaston	(1) David Jones (2) Herberts Park Iron Co.	15 29
89	Highfields	Bilston	(1) J.P. Firmstone (2) W. Riley & Son	26
90	Hill Top	West Bromwich	Bissell, Keay & Bissell	10
91	Hope	Tipton	(1) James & Lowe (2) Hope Iron, Steel & Tinplate Co.	$
92	Horseley	Tipton	(1) Bramah, Cochrane & Deeley Works for sale, 1866	$
93	Horseleyfields (Osier Bed)	Wolverhampton	(1) W. Henderson (2) W. & J.S. Sparrow (3) Osier Bed Iron Co. (4) John Lysaght Ltd	26 24
94	Horseleyfields	Wolverhampton	E.P. & W. Baldwin & Co. Closed 1885	3
95	Hyde★	Kinver	(1) Lee and Bolton (2) H.O. Firmstone Dismantling sale, 1888	20 27
96	Imperial	Wednesbury	John Bagnall & Sons	26
97	Kingshill	Darlaston	(1) Lloyds, Foster & Co. (2) Darlaston Steel & Iron Co. Ltd For sale, 1880	12
98	Lays	Brierley Hill	(1) Hunt & Brown (2) Brown and Freer	33
99	Leabrook (Old Leabrook)	Tipton	(1) Malins & Rawlinson (2) W. Bailey	$

No.	Ironworks	Location	Owners	PF
100	Leabrook	Tipton	(1) John Bagnall & Sons	
			(2) John Bagnall & Sons Ltd	28
101	Leabrook (Patent ironworks)	Tipton	(1) Crudgington and Southall For sale, 1848	
			(2) George Haden & Alfred Hickman	
			(3) Chillington Iron Co.	23
102	Leabrook (New Leabrook)	Tipton	(1) J. Hartland & Co.	
			(2) Solly Brothers	25
			(3) Great Bridge Iron and Steel Co. Ltd For sale, 1873	
103	Level (Old Level)	Brierley Hill	(1) Benjamin Gibbons	
			(2) B. & J. Gibbons	
			(3) Hall, Holcroft & Pearson	
			(4) Henry Hall	18
104	London Works	Oldbury	(1) Oldbury Steel & Sheet Iron Co.	2
			(2) London Works Iron Co.	15
105	Lye Forge★	Lye	John Folkes	1
106	Mars	Wolverhampton	(1) George Adams & Sons	20
			(2) George Adams & Sons Ltd	34
107	Millfields	Bilston	(1) Browning & Jackson	24
			(2) Thomas Rose	20
			(3) Union Coal & Iron Co.	23
			(4) W.H. Sanders	
108	Minerva	Wolverhampton	Isaac Jenks	14
109	Monmore	Wolverhampton	(1) Wright & North	25
			(2) E.T. Wright	
			(3) E.T. Wright & Son	
			(4) Joseph Green, Wright & Alfred Wright trading as E.T. Wright & Son	
110	Monmore Lane	Willenhall	(1) Deakin & Dodd	111
			(2) H. Deakin	
			(3) John Page, William Edwards and James Jellows trading as Monmer Iron Co. Ltd	25
111	Monway (Leabrook)	Wednesbury	(1) William Marshall & Sons	
			(2) Marshall & Mills	12
			(3) J. Marshall For sale, 1883	11

No.	Ironworks	Location	Owners	PF
112	Moxley (Moxley Forge)	Darlaston	(1) E. Creswell & Sons (2) David Rose (3) Southan, Bradley & Co. (4) Moorcroft Sheet Iron Co.	11 10 10
113	Moxley	Darlaston	(1) Thomas Wells (2) Thomas and Charles Wells	25 33
114	Netherton	Dudley	Noah Hingley & Sons	124
115	New Birchills	Walsall	(1) George Jones & Philip Williams (2) John Jones Works split into Birchills, Hall & Staffordshire	$
116	New Golds Hill	West Bromwich	(1) John Bagnall & Sons (2) Bloomer & Davis (3) Thomas Davis (4) Thomas Davis & Co.	12
117	New Lays	Brierley Hill	Brown & Freer	10
118	Nine Locks (Brierley Hill)	Brierley Hill	(1) British Iron Co. (2) New British Iron Co.	18
119	Nine Locks	Brierley Hill	Samuel Newton & Sons	4
120	Oak Farm	Kingswinford	(1) Oak Farm Co. (2) Warden & Williams (3) Sir Stephen Glynne Out of use by 1862	38
121	Old Bradley	Bilston	Enoch Gittings Failed, 1887. Works dismantled, 1891	15
122	Old Church (Tibbington)	Tipton	(1) John & Edmund Walker (2) Walter Robinson & Co. (2) Gospel Oak Iron Co. (3) Gospel Oak Iron and Galvanised Iron and Wire Co. Ltd (4) Tipton Iron & Steel Co. Ltd	27 27 24
123	Old Hill	Old Hill	Noah Hingley & Sons	$
124	Old Park	Wednesbury	(1) Lloyds, Foster & Co. (2) Patent Shaft & Axle Co.	32
125	Oldbury	Oldbury	T. Vernon	3

No.	Ironworks	Location	Owners	PF
126	Park Lane	Oldbury	F.R. Simpson	9
127	Park Lane	Tipton	(1) T. Morris & Son	4
			(2) T. Morris, junior	4
			(3) William Morris	4
128	Pelsall	Pelsall	(1) W.F. Fryer	
			(2) Bloomer & Davis	
			(3) B. Bloomer & Son	26
			(4) Pelsall Coal & Iron Co.	
129	Phoenix	West Bromwich	(1) Malins & Rawlinson	
			(2) Galvanised Iron Co.	30
			(3) Whitehouse & Jeffries	11
			For sale, 1856	
130	Pleck (*Renamed* Newside)	Walsall	(1) Brayford & Lancaster	
			(2) Henry Lancaster & Samuel Jesson	7
			(3) Skelton & Yardley	6
			For sale, 1876, with Walsall IW (qv)	
			(4) John Southan	
			(5) Thomas Southan	
			(6) T. & J.P. Southan	
131	Pot House Bridge	Bilston	(1) Hampton & Brereton	7
			(2) Hampton & Co.	7
132	Providence	Cradley	Joseph Penn & Co.	7
133	Providence (Spon Lane)	West Bromwich	(1) Silvester & Jackson	
			(2) Thomas Silvester & Sons	12
			(3) Bridge, Gill and Bridge	16
134	Rabone Bridge	Smethwick	(1) Job Richards & Co. (Richards & Hill)	14
			(2) Jenkins and Jenkins	
			(3) Rabone Bridge Iron Co.	12
			For sale, 1875	10
135	Regent	Bilston	(1) W. Riley & Son	12
			(2) Ambrose, Beard & Sons	
			(3) Regent Iron Co.	11
			(4) George Onions	11
			(5) Tupper & Co.	
136	Regent Grove (Grove)	Smethwick	(1) John Stones & Son	10
			(2) Sharp & Brown	10
			(3) A Wright	10
			(4) George Beard	10

No.	Ironworks	Location	Owners	PF
			(5) Beard & Eberhart	11
			(6) B.H. Eberhart	15
137	Ridgacre	West Bromwich	(1) Underhill, Whitehouse & Johnson	
			(2) John Whitehouse & Sons (Richard, John & Moses Whitehouse)	12
			(3) S. Whitehouse	20
			(4) Assignees S. Whitehouse & Sons	
			(5) J.T. & W.E. Johnson	
138	Round Oak	Brierley Hill	Earl of Dudley	56
139	Roway	West Bromwich	Edward Page & Sons	20
140	Sheepwash Lane	Tipton	(1) Benjamin Hunt	3
			(2) J. Haines & Co.	10
			(3) Enoch Gittings	
141	Shrubbery	Wolverhampton	(1) Edward & George Thorneycroft	
			(2) G.B. Thorneycroft & Co.	25
			(3) John & Joseph Jones trading as Wolverhampton Corrugated Iron Co.	
			(4) Wolverhampton Corrugated Iron Co. Ltd.	
142	Shut End	Kingswinford	J. Bradley & Co.	34
143	Smethwick (Manchester) (Sandwell)	Smethwick	(1) John Howard Blackwell & George Bennett	
			(2) Joseph Hadley	8
			(3) Hartland & Co.	8
			Later known as 'Sandwell Ironworks'	
			(4) Lones, Vernon and Holden	23
144	Smethwick	Smethwick	(1) J. Hodgetts	3
			(2) John Spittle	3
			(2) Job Legge & Sons	3
145	Spring Vale (Lanesfield)	Bilston	(1) George Jones	
			(2) John Jones & John Murcot	68
			(3) Murcot & Wright	
			(4) George Merriman	
			Part demolished, remainder became Manor Iron & Tinplate Works (Stephen Thompson)	

No.	Ironworks	Location	Owners	PF
146	Staffordshire (New Birchills)	Walsall	(1) Bunce, Jones & Co. (2) Benjamin Bunch & Sons	10 16
147	Stambermill	Stourbridge	Thomas Eveson	$
148	Star	Wolverhampton	(1) Elijah Banner (2) David Jeavons	7 7
149	Stonefield (Factory)	Bilston	(1) Thomas & William Vernon (2) G.H. & A. Hickman (3) Joseph Sankey & Richard Chambers trading as the Bilston Iron Co.	 10 17
150	Stourbridge	Stourbridge	J. Bradley & Co.	23
151	Stour Valley (Spon Lane)	West Bromwich	(1) J. Gregory (2) Weston & Grice (3) Patent Nut and Bolt Co.	6 6 11
152	Stour Valley	Wolverhampton	(1) G.B. Thorneycroft & Co. (2) Shrubbery Steel and Iron Co. (3) Wolverhampton Corrugated Iron Co. Ltd	25
153	Summerhill	Tipton	(1) William & Isaiah Millington (2) Millington & Co. (3) W. Millington & Co. (4) Samuel Lees Millington trading as W. Millington & Co. Dismantled, 1894	 15 16 16
154	Swan Garden	Wolverhampton	(1) G.B. Thorneycroft & Co. (2) John Lysaght Ltd	30 24
155	Swan Village	West Bromwich	D. Hipkins & Son	
156	Swindon★ (Swin)	Swindon	(1) Edward & George Benjamin Thorneycroft (2) Keep & Watkins (3) J. Watkin (4) E.W. and P. Baldwin	 13 24 12
157	Tipton	Tipton	(1) Edward Creswell & Sons (2) Rhos Hall Iron Co. Ltd For sale, 1868	 22 23
158	Tipton	Tipton	E.B. Whitehead	4

No.	Ironworks	Location	Owners	PF
159	Tipton Green (Upper Green Forge)	Tipton	(1) Barrows & Hall (2) W. Barrows & Son	11
160	Tividale	Tipton	Budd & Co.	9
161	Tividale	Tipton	J.O. Haynes	5
162	Toll End (Crown)	Tipton	(1) Horseley Iron Co. (2) Toll End Co. (3) Motteram & Deeley (4) Edward Bayley	9
163	Toll End	Tipton	(1) Kempson & Grafton (2) John Bagnall & Sons (3) John Bagnall & Sons Ltd For sale, 1882	20
164	Victoria	Darlaston	David Rose For sale, 1887	6
165	Victoria	Smethwick	D. & W. Brown	3
166	Victoria	Walsall	(1) Henry Mills (2) Marshall & Mills (3) Henry Mills & Sons (4) Walker Brothers	10 7
167	Victoria	Wednesbury	(1) Edward & George Thorneycroft (2) John Russell For sale, 1854	18
168	Victoria	West Bromwich	D. Hipkins & Sons	12
169	Vulcan	Smethwick	Ebeneezer Parkes	8
170	Walsall *Renamed* Cormorant (Falcon)	Pleck, Walsall	Charles Yardley & Co. For sale, 1876 with Pleck Iw (qv) Harrison, Hopkins & Harrison	12 12
171	Waterloo	West Bromwich	(1) Thomas Johnson Jr (2) Waterloo Iron Co. (3) J.T. & W.E. Johnson	10 12
172	Wedges Mills	Cannock	W. Gilpin Senior & Co.	8
173	Wednesbury Oak	Tipton	P. Williams & Sons	32
174	Wellington	Stourbridge	T. Webb & Sons	11

No.	Ironworks	Location	Owners	PF
175	Wellington (Great Bridge)	West Bromwich	(1) William Cranage & Edward Holden	10
			(2) Lees & Holden	
			(3) Allen & Holden	11
			(4) Wellington Ironworks Co.	
176	Whittington★	Kinver	(1) Hartshorne & Podmore	
			(2) Warden & Williams	
			(3) Whittington Co.	7
			(4) James Williams & Co.	9
			Dismantling Sale 1882	
177	Wilden★	Stourport	(1) Lewty & Co.	
			(2) E.P. & W. Baldwin	7
178	Witton Lane	West Bromwich	(1) Field & Co.	16
			(2) Roberts, Tonks & Co.	13
			(3) Thomas Roberts and Thomas Underhill trading as Roberts & Co.	11
179	Wood Lane	West Bromwich	Bright & Langham	9
180	Woodford	Smethwick	Morewood & Co.	8

REFERENCES AND ACKNOWLEDGEMENTS

Archive Departments and Sources

Birmingham Library Archives – Various deposits, including Lee Crowder MSS and Scott Bateman MSS
Dudley Records Office – Various deposits, including the Earl of Dudley MSS
Lichfield Record Office – Various deposits, including Wills and Bishops

Transcripts

Public Record Office – Reference sources include Birmingham Canal Navigations, Coventry Canal Company and Stourbridge Canal Company
Sandwell Record Office
Stafford Record Office
Walsall Local Studies
Wolverhampton Local Studies

Newspapers

Birmingham Aris's Gazette
Birmingham Daily Post
Birmingham Daily Mail
Birmingham Gazette
Birmingham Journal
Birmingham Mercury
Birmingham Weekly Post
Brierley Hill Advertiser
Dudley Herald
London Gazette
Stafford Advertiser
Walsall Free Press
Warwick & Warwickshire Advertiser
Wolverhampton Chronicle
Worcester Chronicle

Printed Material and Periodicals

The Black Countryman

Colliery Guardian

Dilworth, D., *Tame Mills of Staffordshire*, Phillimore (1976)

Edgbastonia

Edward, I., Iron Production in North Wales, The Canal Era 1795-1850 in *Historical Society Transactions 14, 141-84*, Denbighshire

The Engineer

Hawkes Smith, W., *Birmingham & South Staffordshire* (1838)

Historical Society Transactions

Hunts Mineral Statistics

Iron

Iron & Steel Institute Journal

Journal of the Institution of Civil Engineers

Journal of the Institution of Mechanical Engineers

Kelly's *Trade Directories*, various dates for Staffordshire, Warwickshire and Worcestershire

Mechanics' Journal

Mining Journal

Morton, G.R. and Le Guillou, M., 'Rise and Fall of the South Staffordshire Pig Iron Industry', *British Foundryman* (July 1967)

Morton, G.R. and Gould, J., *Journal of the Iron and Steel Institute* (March 1967)

Pelham, R.A., 'The establishment of the Willoughby Ironworks in North Warwickshire in the 16th century', Vol. IV, *University of Birmingham Historical Journal* (1953–1954)

Scrivenor, H., *History of the Iron Trade*

Special thanks are afforded to:

Dr Peter King, who has provided information about the early iron industry in the district and especially the tables produced at the end of this book.

The author wishes to acknowledge the contributions made and understanding of the trade given by the following:

The late W. Gale; the late S.H.P. Higgins; the late Harold Parsons; the late Ralph Russell; the late Eric Tonks; the late Bill Williams.

John Allen; Jim Andrews; Dr Carl Chinn; Mike Constable.

INDEX